Math
Freak

Jason Earls

Pleroma Publications

ISBN 978-1-312-89874-5

Image information for front cover: Main file found at **Wikimedia Commons** web site: File:27156 FractalBug.jpg; Author: User Åke Persson on sv.wikipedia; Licensing Permission (Reusing the file): **Public domain**; (cover made by Jason Earls using this image).

First Edition
November, 2013
Pleroma Publications

Also by Jason Earls

Numbers for Wittgenstein

The Underground Guitar Handbook

That Man is a Sinner

Computing With Fermat

You Will Be Amazed by the Entertainment

Mathematical Bliss

How to Become a Guitar Player from Hell

Primitive Knife Manual (with Scene Williams)

The Lowbrow Experimental Mathematician

Heartless Bastard In Ecstasy

Concrete Calculator-Word Primes

Red Zen

Morzan the Slayer Must Die

I Sin Every Number

How to Make Your Own Head Explode

*Crazy F*ck*r In Paradise*

A Cringe-Meister in the Bathos-Sphere

Cocoon of Terror

Zombies of the Red Descent

0.13610152128364555667891105120136 15...

Death Knocks

Contents

Math Freak

Weird Problems in Computational Number Theory (With Stories)

by

Jason Earls

Preface

In this book you will find a few fiction stories with math as their primary component, plus a handful of nonfiction articles on number theory. Concerning the short stories, although they are total works of fiction, the mathematics within them is entirely real and accurate. The number theory articles simply comprise a few ideas I thought were interesting at the time; and I hope you too find them interesting.

I'm not sure how many people like reading fiction that contains genuine mathematics. It seems most people do not enjoy it even if it does have authentic number theory results (or *because* it does have them, one of the two). Nevertheless, one of my favorite things to do involves constructing a fictional world with weird characters doing mathematics in unusual ways. Although occasionally I am inspired to write a math article on some particular number theory topic I find intriguing, I like it better when I'm able to incorporate a genuine math idea into a good story. But that's somewhat more difficult to pull off since I don't always have enough of a plot to make a good story to support the mathematical ideas.

In the past I've had a few of my math fiction stories published in web-zines. While some editors were open-minded enough to accept them, more often than not it seemed fiction editors either did not understand the mathematics in the stories and thus felt insulted it was there, or they thought the math was entirely too dry and boring to interest their readers. But not all editors thought this way. One or two really seemed to enjoy my math fiction -- but of course not enough to pay me anything substantial for it! So now I don't bother submitting my fiction anywhere. I would rather just write it to the best of my ability and publish it in one of my own books.

I decided to reprint my story "Mersenne's Mistake," which was originally published in *Bewildering Stories* web-zine and in my book *Mathematical Bliss*. "Mersenne's Mistake" is one of the better math fiction stories I have written over the course of my pseudo-career. For this book, I edited it quite a bit since the original idea was good but the writing was somewhat lacking in places. I wrote the story many years ago when I did not know how to write very well (not that I have improved much).

I hope you like the math fiction stories contained in this book; and if you don't like fiction, perhaps you will at least enjoy a few of the nonfiction math articles. Happy reading.

-Jason Earls, 11/6/2013.

Bar-Hopping and Palindromic Divisors

"Most of number theory has very few 'practical' applications.
That does not reduce its importance, and if anything it
enhances its fascination."
-C. Stanley Ogilvy and John T. Anderson

"Intellectuals solve problems, geniuses prevent them."
-Albert Einstein

I have a certain friend that I've known for many years, let's call him Fred (not his real name). Fred has a good mind for mechanical things, he works for a company whose central headquarters are in Canada and Fred's primary job is to invent, design, and build various machines for foundries. Frequently Fred's boss will send him to one of the foundries in Canada where he has to set up a new machine that he has designed, then Fred must test everything about the machine, make sure it's functioning correctly, and train employees in the foundry to use it properly.

While on these visits to Canada, after work Fred usually becomes quite bored, and more often than not he ends up in bars drinking copious amounts of alcoholic beverages. Fred is married and a thoroughly moral man so he does not try to pick up women in the bars, he simply drinks and hangs out with people and attempts to relieve his intense loneliness of being so far away from home. Fred is half-extrovert and half-introvert and somewhat of a people person when his mind is not absorbed by

Math Freak
by Jason Earls

designing and building foundry machines. But lately Fred has realized that he has been drinking far too much (and spending too much money) in Canadian bars and he does not want to turn into a full-blown alcoholic. So all of that is Fred's basic background, but there is plenty more to learn about him, which you'll soon find out below.

Recently I ran into Fred on the street and he asked me what I'd been up to lately. I mentioned that I had been intrigued with number theory. Usually people are not interested in any type of mathematics, but Fred was a pretty technical person so I thought I would tell him the truth in case he was genuinely interested. "How is number theory different from math?" Fred asked, and I could tell he really wanted to know, that he wasn't just being polite. I did my best to explain how number theory differed from the traditional math problems teachers assigned us in school, and Fred followed along, really concentrating on what I was saying. Next I explained a few recent results I'd found with computer searches and he replied to each with enthusiastic expressions such as, "Neat!" or "Cool!" I never knew Fred was interesting in mathematical concepts, always believing him to be more of a hands-on mechanical person. It's very rare to find anyone who can really appreciate results in pure number theory, since I believe it takes a certain kind of person with a certain kind of mind that prefers high levels of abstraction. So after discussing number theory for a few minutes, Fred and I chatted about a few personal topics, then we bid each other goodbye and good luck.

I thought Fred would soon completely forget about the

Math Freak
by Jason Earls

number theory subjects we had talked about, since so many people just talk for entertainment and take nothing anyone says to heart. It's as if talking is some kind of passive recreation and nothing about it is serious and you're just supposed to immediately forget what was said and no one cares what the other person was talking about in the first place. But that wasn't the case this time with meeting Fred on the street.

A few months passed and I was in the grocery store picking up some grapes and milk when I ran into Fred again. He seemed excited to see me. "Do you have time to follow me to my house?" he asked in a visibly excited state.

"Oh sure. What for, Fred?" I said.

"I want you to see some of the math stuff I've been working on."

"Math stuff?" I said. "I thought you were an engineer. You're still designing and building machines for foundries, right?"

"Yeah, of course I still do that," he said. "But I've been working on number theory lately and I want to get your opinion on some results I've found."

I was shocked. I thought the interest Fred demonstrated in our last conversation would quickly fizzle out. "Yeah, I'll take a look," I said.

I got into my Oldsmobile Intrigue and followed his

Math Freak
by Jason Earls

pickup over to his house. I walked up the sidewalk but Fred waved me over to one side and I followed him into his back yard.

"Where are we going?" I said.

"I have a little shed back here where I work. It's so I can get away from the wife and kids to concentrate better on my math work."

Fred led me over to a small tin building which turned out to be a little grungy tool shed, stuffed with common garage tools and lawn equipment. He unlocked the small rickety door and we stepped inside. The place smelled strongly of motor oil. There was a small work bench Fred was using for a desk, a metal folding chair, and a small fan hanging from the ceiling in a precarious way. Fred had turned the little shed into his own personal math research center. Many pads and pens were on the desk with a few books and some old calculators. I didn't see a computer anywhere, which was my favorite way to probe the integers.

"No computer?" I said.

"Nah, I don't know how to use a computer for math or number theory," he said. "I only know how to use the programs they trained me in at work to design certain parts of the foundry machines I build, and I use my work computer for that. I don't have a laptop at home or even a smart phone. I have done some computer programming before, but I'm not very good at it."

Math Freak
by Jason Earls

Fred paused and looked around with his hands in his pockets, a slightly embarrassed grimace on his face, which I guessed was from him choosing to do mathematical research in a small dirty tool shed.

"After you showed me those results awhile back," Fred continued. "I went and borrowed some books on number theory from the library and started doing some rough calculations. Soon I found some results of my own, which is what I wanted to show you. I became interested in palindromes. I have a few results and I want to see if you think they're significant in some way. Here, take a look at this."

He reached over and grabbed one of about five school notebooks and flipped through it. He opened it toward the middle and pointed to some calculations. The pages seemed completely filled with numbers and symbols I didn't quite recognize. Fred seemed to have made up his own notation for certain concepts and I couldn't understand any of it. But there were plenty of numbers and patterns that he'd written out and those I could easily understand.

"Here is one of my best pages of results," he said. "Does this seem significant?"

I looked at the page and recognized some palindromic numbers (those that read the same way forward and backward), but the symbols next to them I couldn't comprehend -- his personal notation. "What does this

Math Freak
by Jason Earls

mean?" I said, pointing to the number 2030202 with some weird symbols surrounding it.

"Well, let me break it down for you. This number 2030202 is really special. It isn't a palindrome, yet it has a lot of divisors, but only one of them is a palindromic divisor with a lot of digits, while the other palindromic divisors are trivial in the sense that they have a low amount of digits. 2030202 is a very special number."

I realized I would have to write down some of his results and check them with my computer at home, but for the moment I took his word for it. "How did you discover this fact?" I said. "Do you have a computer?"

"No," Fred said, shaking his head and taking back his notebook. "I told you, I don't know how to program this kind of stuff with a computer. I only have a few old calculators over there, but they don't handle large integers since their displays only show a few digits. The majority of my calculations are done with pen and paper, that's how I found 2030202. Let me give you an example of how I calculate."

Fred tried to explain his personal method of computing but it was extremely weird and he was using strange words that didn't make sense to me. He had indeed invented his own terminology and made-up phrases for common number theory operations. Nevertheless, I made a strong effort to follow what he was saying. His results did seem genuine, and if they really were accurate it was amazing he was able to find them with only pen, paper, and calculator.

Math Freak
by Jason Earls

"Here, let me show you this other notebook. I have some results on Sierpinski divisors you might think are interesting."

"Sierpinski divisors?" I asked.

"Isn't that a well-known set of divisors?"

"I've heard of Sierpinski numbers, but not divisors. Anyway, this palindromic result is enough for me right now. Let me borrow a sheet of paper and write down the number you showed me along with its properties and I'll check it at home, then I'll get back to you and look at your other results. This is great so far, Fred. Really cool stuff. I'll see you later."

I didn't mean to be rude but I was eager to get home and check his results for accuracy. After jotting down the number and a few notes I practically ran out of the grungy tool shed over to my car. I drove home and rushed into my house, went over to my computer and immediately began exploring Fred's palindromic result. To be sure I was following all his claims, I went down each one in check-list form:

1. He said 2030202 is not a palindrome. Correct.

2. He said it has a lot of divisors. I computed its number of divisors with numdiv(2030202) = 36, which I guess is a lot for a guy using only a calculator. Check.

Math Freak
by Jason Earls

3. He claimed 2030202 has one large palindromic divisor, meaning it has a lot of digits. I computed its full list of divisors: [1, 2, 3, 6, 9, 18, 43, 61, 86, 122, 129, 183, 258, 366, 387, 549, 774, 1098, 1849, 2623, 3698, 5246, 5547, 7869, 11094, 15738, 16641, 23607, 33282, 47214, 112789, 225578, 338367, 676734, 1015101, 2030202], indeed 1015101 is a large palindromic divisor. Check.

4. He said its other palindromic divisors were trivial in the sense that they have a low amount of digits. 1, 2, 3, 6, 9, were the only other palindromic divisors and they had only one digit, so that's why he deemed them trivial. Check.

All of Fred's results checked out. 2030202 was a pretty special number indeed. Wow, he had found all this with pen, pad, and a calculator with barely any digital display. Amazing. Not only was the man skilled at designing and building foundry machines, he was also skilled with numbers. Holy smokes. I knew I had to return and visit Fred in his grungy tool shed the next day to let him know everything he claimed about 2030202 was indeed true.

I drove to his house around six in the evening when I knew he would be off work. I knocked on the front door. His wife answered and said Fred was out in his shed working on his "math crap." Obviously she didn't like his new hobby, or the time he spent engaged in it.

I went around to the back yard. There was a tiny window on the door to his shed and I couldn't keep from peeking inside to see if Fred was engaged in his work. First I saw

Math Freak
by Jason Earls

only the abundance of tools on the walls, then hunched over his small dirty work table surrounded by loose papers and books, the small white fan hanging from the ceiling blowing away, there sat Fred furiously punching buttons on one of his calculators.

I knocked on the shed door lightly. Fred lurched up with a bewildered look on his face as if he'd been broken from a deep reverie of concentration. He recognized me and a huge smile broke out on his face. "Come in! Come on in!" he yelled, rising from his work table. "I've made a new discovery and I can't wait to show it to you!"

"Hey Fred," I said opening the door and stepping in. "I checked your results and everything was perfectly correct. No mistakes whatsoever."

"Great. Wait till you see what else I found."

"Have you proved something significant?" I said, looking at a rusty spade leaning against his bench about to fall on him at any second.

"Proved? What does that mean?"

"Oh, never mind. Let's hear about your latest discovery."

"Well, at first I was working on a sequence of the first numbers that have only one nontrivial divisor of length n. For example, 143 is the first non-palindromic number that has only one 2-digit palindromic divisor. Then 1313 is the first non-palindromic number with exactly one 3-digit palindromic divisor (other than the trivial number

Math Freak
by Jason Earls

1). Then the definition gets a little vague and I'm not sure how to continue since it's hard to find numbers that are not palindromic but still have one single very large palindromic divisor but without any nontrivial divisors lower. What I mean by that is, look at this full list of divisors for 215402."

He thrust his notebook at me and I saw: [1, 2, 11, 22, 9791, 19582, 107701, 215402].

"The number is non-palindromic and it has 107701 as a divisor, but it also has the -- in my opinion trivial -- divisors of 1, 2, 11, and 22. So my definition starts to get a little vague about what exactly non-trivial divisors are."

"Wow, that's great," I said, genuinely surprised. "I can't believe you've made all of these discoveries in such a short amount of time." I sat down on a couple of old tires stacked up, the top one completely bald with no tread.

"Yeah well I cured my drinking problem. Whenever I go to Canada now I don't go to the bars anymore. I stay in my hotel and work on number theory. It's so damn fun working on these little problems. I'm sure real mathematicians would think they are trivial, but to me they're amazing. Now that I've found computational number theory I don't do all of that partying and waste so much of my money. I just work during the day and stay in my hotel all night tripping out on math and numbers. In my hotel room I always have a notebook and some pens and I take along a few books to stimulate thought. Mainly I just try to read about interesting math problems that I

Math Freak
by Jason Earls

find interesting and if I feel I can do some calculations I start working away and investigating properties until every once in awhile I make a little discovery. It's weird. I've never had any previous interest in math at all until the day you told me about your number theory results. Now it has developed into this obsessive thing that's taking over my life."

"That's outstanding, Fred" I said. "I mean, I hope it doesn't get out of hand. But your work so far is incredible. So what's the new discovery you mentioned?"

"Well, I'm still working on the palindromic divisors thing. Check out this number and its divisors." He held out a crumpled sheet of yellow notebook paper with 219802 and all of its divisors written out: [1, 2, 11, 22, 97, 103, 194, 206, 1067, 1133, 2134, 2266, 9991, 19982, 109901, 219802].

"I'm still looking for the largest non-palindromic number that has one very large palindromic divisor, but the thing I don't like is that the small palindromic divisors are at the beginning and the biggest palindromic divisor is always the same as the largest proper divisor. See how this one has 5 palindromic divisors, the first 4 of which only have 1 or 2 digits, while the largest has 6 digits? I wish I could find one that has its one and only largest palindromic divisor more toward the middle. Why do they always have to be the damned largest proper divisor?"

I was glad he had been reading more number theory

by Jason Earls

books and was using terminology I could actually understand. "I don't know. But it's amazing you're using only a calculator to find these numbers. Do you have an algorithm for finding them?"

He wrinkled his face. "What's an algorithm?" he said in all seriousness.

"Never mind, I'll explain later. What else have you found?"

Fred leaned over and grabbed a couple more sheets of crumpled paper off his desk and began pointing out things. "Another one similar to the first is 2014002 which has these divisors: [1, 2, 3, 6, 9, 18, 41, 82, 123, 246, 369, 738, 2729, 5458, 8187, 16374, 24561, 49122, 111889, 223778, 335667, 671334, 1007001, 2014002]. But see how it's following almost the exact same pattern as the previous number? I don't like that very much. I want them to be more unpredictable. I want to find something really surprising. So next I tried finding some non-palindromes such that their largest palindromic divisor was not also their largest proper divisor. And this is the number that led me to my main discovery that I wanted to tell you about. **1300013**. Here are its divisors:

[1, 11, 13, 143, 9091, 100001, 118183, 1300013]

"Look at the basic pattern. It is non-palindromic yet it has only 3 palindromic divisors, the largest of which is NOT its largest proper divisor. But look closely at the form of the number? It is $p...000...p$ where p is a prime and the rest are zeros! Holy cow! When I recognized that I knew I was on to something. I did more calculations and this form of number demonstrates most of the properties I've been curious about. It involves numbers of the form:

$$A(n) = \text{prime}(n) * 10^{\wedge}n + \text{prime}(n)$$

"I'm trying to determine if the pattern of one palindromic divisor 2nd from the largest, with very small initial palindromic divisors, will continue to hold. For the ones I'm able to calculate, the pattern is still there. But I'm having trouble calculating these numbers since they're so large. What exactly I'm trying to find are numbers that are non-palindromic but with a really high amount of regular divisors, but only one very large palindromic divisor, and any other palindromic divisors it has are trivial with a very low amount of digits."

I sat there with my mouth open. I couldn't believe he was making these discoveries without the use of a computer and after such a short amount of time doing number theory. "Wow, I think I can help you, Fred. Let me jot down a few things." I grabbed pen and paper and made some notes. "I'll take this home, program whatever I can, and see what I can find. I really like your work so far. It's excellent in every way. I'll be back in a few days and let you know what I found. Gotta go. See ya soon."

Again I was extremely excited to rush home and check Fred's results. I almost tripped and fell over one of those old-fashioned push reel mowers with no engine on the way out of his yard, but I was in such a hurry to explore his discoveries I didn't care that I'd almost broken my leg. I tucked my little piece of paper in my pocket and drove

Math Freak
by Jason Earls

home. I thought about how much Fred seemed to like palindromes. I did too and so do most math amateurs when they begin messing around with number theory. Funny little visual properties of integers are neat, even though they're the type of things that usually sicken real mathematicians. I didn't care, I was going to explore them. I pulled into my driveway and went inside the house and straight over to my computer. I began testing Fred's claims just as I'd done previously. First I put in Fred's main function: **$A(n)=prime(n) * 10^n + prime(n)$**, then typed up a simple "for loop" and pressed Enter. Here is what the numbers look like:

22,
303,
5005,
70007,
1100011,
13000013,
170000017,
1900000019,
23000000023,
290000000029,
3100000000031,
37000000000037,
410000000000041,
4300000000000043,
47000000000000047,
530000000000000053,
5900000000000000059,
61000000000000000061,
6700000000000000000067,

Math Freak
by Jason Earls

71000000000000000000071,
73000000000000000000073,
79000000000000000000079,
83000000000000000000083,
89000000000000000000089,
97000000000000000000097,
101000000000000000000000101,

Pretty cool. Next I had to write a function that returned
the number of palindromic divisors of a given number,
and this took me about ten minutes. I wrote a small
program that computed the number of palindromic
divisors of A(n) and compared it to the number of regular
divisors of A(n). Here is the data (*format is "n:number of
palindromic divisors:number of regular divisors"*):

**1:4:4, 2:4:4, 3:8:16, 4:5:8, 5:5:6, 6:3:8, 7:3:8,
8:3:8, 9:8:64, 10:3:16, 11:5:48, 12:3:16, 13:4:16,
14:3:32, 15:8:256, 16:3:64, 17:3:32, 18:4:16,
19:4:8, 20:3:32, 21:12:384, 22:3:32, 23:3:64,
24:3:16, 25:4:64, 26:5:12, 27:7:256, 28:3:32,
29:3:16, 30:5:256, 31:3:8, 32:4:32, 33:16:768,
34:3:32, 35:5:64, 36:8:128, 37:3:32, 38:3:16,
39:9:1536, 40:3:32, 41:3:16, 42:10:256, 43:5:32,
44:3:32, 45:10:8192, 46:3:32, 47:3:32, 48:4:1024,
49:4:32, 50:4:256, 51:8:512, 52:3:32, 53:3:8,
54:5:128, 55:9:1536, 56:3:64, 57:8:128, 58:3:64,
59:3:32, 60:5:128, 61:4:16, 62:3:16, 63:13:6144,
64:2:16, 65:15:128, 66:5:512, 67:3:8, 68:3:32,
69:8:2048, 70:5:4096, 71:6:16, 72:4:64, ...**

From examining the outstanding cases in the data above,
it appeared Fred had really hit upon something

Math Freak
by Jason Earls

interesting here. He said he wanted to find values with a high amount of regular divisors, and only one palindromic divisor that was not its largest proper divisor. The greatest example I had found so far was A(45), which has 8192 total divisors and only 10 of them are palindromic! Incredible. Notice that many of the values have only 3 palindromic divisors. I wonder if there will be infinitely many of those? Also look at the values for A(64), it only has 2 palindromic divisors! (but 16 regular which is a low amount compared to the others). Next I wanted to print out some examples of the A(n) function with a lower amount of total divisors, which are more manageable to examine, such as the divisors for A(9) below (notice I put '*'s surrounding each of its 8 palindromes):

[*1*, *7*, *11*, 13, 19, 23, *77*, 91, 133, 143, *161*, 209, 247, 253, 299, 437, *1001*, 1463, 1729, *1771*, 2093, 2717, 3059, 3289, 4807, 5681, 19019, 23023, 33649, 39767, 52579, 62491, 368053, 437437, 578369, 683527, 999001, 1209317, 4048583, 4784689, 6993007, 7518797, 8465219, 10989011, 12987013, 13302487, 15721121, 22977023, 52631579, 76923077, 90909091, 93117409, 110047847, 142857143, 160839161, 172932331, 252747253, 298701299, *1000000001*, 1210526317, 1769230771, 2090909093, 3285714289, 23000000023]

The patterns displayed in some of the divisors were indeed striking. As previously mentioned, A(64) was particularly amazing since it has only 2 palindromic divisors: 1, and the huge 1000...0001 divisor which seems to be common to all of the A(n) numbers. But I would need to rigorously prove that.

[1, 311, 1265011073, 393418443703, 1534316818888913781836 9, 4771725306744521861512759, 19409277653846114909659847799937, 515217525265213267447869906815873, 60362853503461417369042126657804 07, 1602326503574813261762875410197 36503, 6517558744641520450280658823855575171617 2 9,2026960769583512860037284894219083878 3 7297719,79050691440074058545414803653659 4 016043051664260017113 7,24584765037863032 2 0762400393628807389893890675848653223 60 7, 10000000000000000000000000000000000 00000000000000000000000000000000001, 31100000000000000000000000000000000000 00000000000000000000000000000311]

Phenomenal. The display of huge divisors with only two of them being palindromic, it's insane! This seemed like one of the pure specimens Fred had been looking for. Would this ever happen again? What an incredible result. I decided to run a computer program to search for another A(n) value with only 2 palindromic divisors. I was determined to continue the search for months if I had to.

After running the computations above, I printed out the main results so I could show them to Fred. But I wanted to wait a few days before visiting him again so his wife

Math Freak
by Jason Earls

wouldn't get tired of seeing me. After three days I couldn't take it any longer. I went to his tool shed on a Sunday afternoon and found him at his work bench calculating away. I knocked and barged in immediately and handed Fred the printouts of what I had found. His face lit up. He had never seen any computer data on his problem before. He seemed impressed, but not totally blown away, as if he had already seen the walls of divisors for the numbers in his mind and this was only confirming what he had already seen using only his primitive calculator. It was then that I suspected Fred must have savant-like calculating abilities and only rarely would he need access to a computer.

He examined the data for several minutes without speaking. Finally he said: "I like it. I like this a lot. But I am slightly disappointed by one thing. I don't like that the largest palindromic divisor always seems to be of the form 1...000...1. I want the biggest palindrome to occasionally be different from that."

"You can't always get what you want, Fred. Isn't that a *Rolling Stones* song? Never mind. Plus, we haven't proved that the largest palindromic divisor will *always* be of the form 1...000...1. There may be some that don't follow that pattern. That's why in math you have to prove stuff."

Fred scratched his chin. "But after looking at your data, I can see in my mind that they will always have the 1...000...1 palindromic divisor as the main one."

"You can see it in your mind?" I said.

Math Freak
by Jason Earls

"Of course. And they will never be different. Once I see some initial data on a problem, the natural structure underlying it just forms in my mind and I know certain patterns will continue to hold no matter what. I have never been wrong about my hunches either. That's the main skill I have that allows me to do extreme calculations without a computer."

Hmm... Seems he did have savant-like abilities. "Okay, that's great, Fred. But did you the notice the data on $A(64)$? Isn't it amazing that it has only 2 palindromic divisors of 1 and the huge 1000...0001? Do you think we will ever find another $A(n)$ value like that?"

"That is really interesting. I do think there will be more of those. There must be. That's definitely a type of number I had in mind when I first began investigating this problem. $A(64)$ is probably the most interesting case we've found so far. But did you notice something else? Sometimes numbers of the form $p...000...p$ will be palindromes."

"Of course, when p itself is a palindrome then $A(p)$ will be a palindrome and that goes against your original definition."

"And you didn't eliminate those from your computer search, did you?"

Before I could respond, we heard a door slam loudly that startled both of us. We quickly turned and saw Fred's son A.K. walking through the yard. He came out to the tool

Math Freak
by Jason Earls

shed and said: "Hey Dad, Mom wants you to come in for dinner."

"All right, I'll be in in a minute." Fred said without looking at his son.

Out of nowhere, A.K exploded. "COME IN NOW! STOP WORKING ON THIS DAMN MATH GARBAGE ALL DAY AND ALL NIGHT! IT'S NOT RIGHT! YOU'RE NEGLECTING YOUR FAMILY! WHAT KIND OF FATHER ARE YOU ANYWAY!" A.K. broke down and started sobbing.

"YOU SHUT UP, A.K.!" Fred yelled. "I DON'T NEGLECT YOU! I PUT BEANS IN YOUR BELLY AND CLOTHES ON YOUR BACK, DON'T I?"

"THAT'S ABOUT ALL YOU DO, YOU BASTARD!"

"WHAT! YOU CAN'T TALK TO ME LIKE THAT!" Fred took off running toward his son and I knew I had to get the hell out of there. We had done enough computational number theory for one day.

I drove home thinking about what I would do next. I would continue the search for another $A(n)$ value that had only two palindromic divisors, while working on other number theory problems of my own. Maybe I would try to prove one or two things about Fred's $A(n)$ numbers. I also decided I would stay away from Fred's little grungy tool shed for awhile, since he was having far too many family problems for my taste. But I knew I would see him again soon, partly to monitor his amazing progress, and also because I would be really curious about any new results he had found.

Questions

1. Do you agree with Fred that there will definitely be another $A(n)$ value that has only two palindromic divisors?

2. Can you compute more divisors of the $A(n)$ function and find anything interesting?

3. What other functions can you find that produce numbers that better fit Fred's basic definition of the palindromic results he was looking for in the story?

Math Freak
by Jason Earls

Variations of Brocard's Conjecture

"The essence of mathematics is its freedom."
-Geoge Cantor

"I have created a new universe from nothing."
-Janos Bolyai

Brocard's conjecture states that between the squares of any two consecutive prime numbers, there are at least 4 primes. For example, between $5^2=25$ and $7^2=49$ there are 6 primes: (29, 31, 37, 41, 43, 47).

Computational evidence suggests that Brocard's conjecture is almost certainly true: between the squares of the 99th and 100th primes for example, there are no less than 1513 primes!

Let's create a variation of Brocard's conjecture with semiprimes and see what happens. A semiprime is a number that's the product of exactly two primes, e.g. $15=5*3$ and $123467 = 311*397$. So between the squares of any two consecutive semiprime numbers, how many semiprimes can we find? A lot. Said sequence begins like so:

7, 13, 8, 28, 13, 60, 14, 41, 12, 118, 20, 20,

For example, the first and second semiprimes are 4 and 6, which become 16 and 36 when squared, and between those values are 7 semiprimes: (21, 22, 25, 26, 33, 34, 35), which is the first term in the sequence above.

After running many computational experiments it appears we can go so far as to state this:

Conjecture: There are at least 7 semiprimes between consecutive squared values of semiprimes.

More computational evidence for the conjecture: For $n>=253$ it seems there are always at least 100 semiprimes between successive squared semiprimes; and for $n>=2845$ there appears to be at least 1000 semiprimes between successive squared semiprimes. Holy smokes, that's a lot of semiprimes my friends!

Can we make another variation of Brocard's conjecture? Damn straight we can. Let's try brilliant numbers between successive squared semiprimes and run a few quick and dirty computations.

First, a brilliant number is defined as a semiprime whose prime factors have the same number of decimal digits. For example, $8973037=1009*8893$ is a brilliant number since each of its prime factors have exactly 4 digits.

The sequence: "number of brilliants between the squares of any two consecutive semiprimes" begins like so:

3, 1, 0, 4, 2, 13, 3, 10, 4, 35, 3, 5, 14, 4, 27,
10, 9, 14, 7, 2, 13, 9, 9, 11, 6, 7, 4, 1 ...

Now, based on this highly limited sample of data, if we

wanted to "jump the gun" and make a conjecture that for all $n >= 10$ there will always be at least one brilliant number between successive squared semiprimes, **WE WOULD BE VERY WRONG INDEED.**

After running a bunch of computer experiments to see where it might be "safe" in the number line to try to make a cool yet confident conjecture that there will always be at least one brilliant between consecutive semiprimes, my program reached the 2000s range and for an instant I thought, "Okay dude, it's probably safe to make the conjecture now." But then still observing the computer data and letting the program climb into the 3000s, 4000s, and 5000s, where the amount of brilliants was hovering around the 100 range, gradually I noticed that in the 7000s the values began to get a little too scarce for comfort; until finally I was completely shocked to find the wall of data below (pay attention to the number on the far right) - *[format below is consecutive semis: their square values: then the count of brilliant numbers between those squares]*:

```
9598:9599:92121604:92140801,count:12
9626:9627:92659876:92679129,count:8
9662:9663:93354244:93373569,count:7
9753:9754:95121009:95140516,count:7
9754:9755:95140516:95160025,count:5
9777:9778:95589729:95609284,count:5
9862:9863:97259044:97278769,count:2
9865:9866:97318225:97337956,count:2
9902:9903:98049604:98069409,count:4
9913:9914:98267569:98287396,count:1
9934:9935:98684356:98704225,count:1
9937:9938:98743969:98763844,count:0
9938:9939:98763844:98783721,count:1
```

Math Freak
by Jason Earls

9985:9986:99700225:99720196,count:0
9986:9987:99720196:99740169,count:0
9997:9998:99940009:99960004,count:0
10018:10019:100360324:100380361,count:0
10021:10022:100420441:100440484,count:1
10041:10042:100821681:100841764,count:0
10077:10078:101545929:101566084,count:0
10117:10118:102353689:102373924,count:2
10118:10119:102373924:102394161,count:4
10173:10174:103489929:103510276,count:1
10201:10202:104060401:104080804,count:10
10213:10214:104305369:104325796,count:6

What the hell! We ran into a **BUNCH OF FRICKIN' ZEROES IN THE 9000S and 10,000S RANGE!** Which means no brilliants there at all. Thus it appears the mighty brilliant numbers are not so easily tamed after all! Nobody can tame the brilliant numbers. Not ever! **THEY ARE SIMPLY TOO INCREDIBLE AND AWESOME AND WILD.**

After my excitement dwindled down, I pulled myself together and realized there may even be more zeros in that particular range than those shown above since in my computer program I only tested the n and $n+1$ values such that they are both consecutive semiprimes, which leaves other semiprime ranges out of the testing.

So concerning any conjectures that could be made for the brilliant version of Brocard's conjecture, I am afraid **none can be made!** The brilliant numbers are simply **too defiant and unpredictable.** Which is why they are by far my favorite class of numbers.

Pell's Equation and Jesus's Number

"Mathematics is not best learned passively; you don't sop it up like a romance novel. You've got to go out to it, aggressive and alert, like a chess master pursuing checkmate."
-Robert Kanigel

"For out of fear and need each religion is born, creeping into existence on the byways of reason."
-Nietzsche

"Even the truth, when believed, is a lie. You must experience the truth, not believe it."
-Werner Erhard

Pell's equation and the number 888. What do they have in common? Not much. But if you have a strange excitable mind capable of contriving bizarre and vicious mathematical objects, they can be forced to have at least one thing in common.

Firstly, we need to explain Pell's equation. According to James J. Tattersall's book *Elementary Number Theory in Nine Chapters*, Pellian equations should actually be called "Fermat's equations" since the brilliant amateur number theorist Pierre de Fermat was the first person to investigate solutions to:

$$x^2 - d*y^2 = 1$$

where x is not equal to 1 and y is not equal to 0.

Math Freak
by Jason Earls

Although it won't be explained in detail here, Archimedes once constructed a problem to taunt a fellow mathematician that concerned cattle which involved solving the intense Pellian equation:

$$x^2 - 4729494^*y^2 = 1$$

and Archimedes's problem went unsolved for almost 2000 years! While mathematical super-genius Leonard Euler used Pellian equations to determine whether numbers are both square and triangular. In this article however, we will be concerned with another property of Pellian equations...

Now let's explain Jesus's number: it's simply the positive integer 888. Why? Because if you write the name 'Jesus' in Greek, then sum the letters using their Gematria values, the result is 888, like so:

Jesus (ΙΗΣΟΥΣ)
Iota = I =10
Eta = H = 8
Sigma = Σ = 200
Omicron = O = 70
Upsilon = Y = 400
Sigma = Σ = 200

Total = 888

Quite a nice little numerology curio, is it not? I know what you're thinking... numerology is nonsensical garbage, but in small doses it really isn't so despicable. Now, what would happen if we combined Jesus's number into a Pellian equation? Could there be any solutions to:

Math Freak
by Jason Earls

$$x^2 - 888*y^2 = 1$$

Of course there are. All it takes is a moderately-powerful spare ghetto computer sitting in your damp basement that no one is using to find a couple of pesky solutions. So haul your rump down there and fire up that number cruncher and begin programming away at once (just kidding)! Here are two solutions:

$$149^2 - 888*5^2 = 1$$
$$44401^2 - 888*1490^2 = 1$$

Do you think we are the first people ever to lay eyes on these specific Pellian equations? Probably not. Pellian equations have been around for many centuries and various mathematicians have even gone so far as to enumerate charts of solutions for the first few thousand integers or so. Therefore I seriously doubt we're the first. But I didn't scour the internet looking for previous solutions or anything, so I don't really know. I simply programmed a few lines of code, basically two simple "for loops," and let the computer run amok for a few hours until the aforementioned solutions popped up. **BUT NOTICE THAT ONLY TWO MEASLY SOLUTIONS WERE FOUND.** What a crock!

Anyway... Now that those solutions have been discovered, we can demonstrate yet another property of Pellian equations. Their other practical purpose is to give approximations to sqrt(d), which in our case is of course Jesus's number 888. Observe:

Sqrt(888) =
29.79932885150267943866363209224...

Math Freak
by Jason Earls

First approximation:
149/5 =
29.800000000000000...

Second approximation:
44401/1490 =
29.799328859060402684563758389 2617...

The second fraction agrees to eight digits after the decimal point! See how we used the *x* and *y* values in the Jesus number Pellian equation to find the approximate values of sqrt(888)? Isn't it sweet that they agree for the first eight digits?

I wonder if Jesus Christ likes mathematics? (No disrespect intended, I'm just innocently speculating here.) Was there even a subject known as mathematics during Christ's earthly lifetime? I wonder if Jesus knew how to take square roots? Or if he liked prime numbers at all?

If I'm not mistaken, after Jesus Christ "rose to power," I think he became omniscient the same way that his father, **God**, is totally omniscient. Which means that now he has all the mathematical knowledge and ability anyone could ever want. Wouldn't it be fantastic to have a divine meeting with Jesus Christ and ask him about the various unsolved problems in number theory? I bet Jesus even knew all about the full proof of Fermat's Last Theorem before Andrew Wiles officially proved it in 1994. Except Jesus probably knew Fermat's **original proof** that he (famously) did not have room enough to give in the small margin of his copy of Bachet's *Arithmetica*. (Fermat's original proof is probably rotting away in a forgotten trunk somewhere in France right now...) Wouldn't it be

Math Freak
by Jason Earls

incredible to hand Jesus Christ a piece of notebook paper
and watch him write down intense proofs to a few
unsolved problems in number theory? Perhaps he could
give the proof that odd perfect numbers do not exist... Or
the proof that there are infinitely many twin primes... Or
an impossibility proof for Goldbach's Conjecture (every
even number greater than 2 is always the sum of at least
two primes); that is, that the laws of our number system
do not allow that problem to be proved either true or
false! Perhaps Jesus could even give me a simple proof
that there are infinitely many primes of the form $n^2 + 1$ ...

I'll bet Jesus could prove each of the aforementioned
results using only high school mathematics that almost
everybody in the world can easily understand. Just as
Euclid had a short and brilliant proof for the infinitude of
primes, I'll bet Jesus could write down a single page of
brilliant logic for each problem above that would
rigorously prove it. Wow, wouldn't it be totally mind-
blowing to watch Jesus Christ writing down these
amazing proofs right before our eyes? And after he did so,
I would try to memorize each one so I would have them in
my mind forever. Except I wouldn't be able to reveal
them to the world because then I'd have to say Jesus
Christ is the true author and not myself. I couldn't take
credit for His work. That would be cheating and lying and
Jesus has no tolerance for that sort of immoral behavior.
Maybe Jesus Christ could even show me proofs of
number theory results that are **so advanced** us mere
mortals here on Earth don't even know they exist yet.
Man, it would be so incredible to see those new problems
with their proofs that it's giving me chills just thinking
about it...

Questions

1. Do you think Jesus Christ would like this math paper? Do you think he is omniscient and possesses more mathematical ability than any other person in history?

2. Can you find more solutions to the Jesus Pellian equation discussed above? Can you give better approximations to **sqrt(888)** in another way? Is it possible to use continued fractions to find solutions to the Jesus Pellian equation? Do you know how to do it?

3. If you could push a button and raise your I.Q. by enough points to solve any mathematical problem you desired, would you do it? By how much would you have to raise your I.Q. to solve any number theory problem in the world? 100, 200, ... maybe 1000 points? What would be the consequences of a person having an I.Q. that high? What do you think Jesus's I.Q. is?

4. Do you think Jesus Christ would consider reading and studying this math paper to be a sin? Or do you think he encourages many different forms of intellectual curiosity and avidity?

On Reversible Prime Numbers (Base-10)

"I have recently carried out some numerical studies of the distribution of prime numbers into certain arithmetic progressions, on which I hope to report fully in due course. Some of the results are so surprising, however, as to be worth making the subject of a separate note."
-John Leech

"The true work of art is but a shadow of the divine perfection."
-Michelangelo

"The logic of the world, which is shown in tautologies by the propositions of logic, is shown in equations by mathematics."
-Ludwig Wittgenstein

Dr. Clifford Pickover recently posted this math curiosity on his Twitter page (I'm paraphrasing):

1111111999999 is a prime number; while its reversal is also prime: 9999991111111.

Upon reading it, I immediately wanted to run a computer search to find an even larger prime number of the same type. But first I had to create a function that produced increasingly larger numbers that displayed the same base-10 digital pattern. After a few minutes of tinkering, here is what I came up with:

$$z(n) = (10^n - 1)/9 * 10^{n-1} + (9*(10^{n-1} - 1)/9)$$

And testing it produced these numbers:

Math Freak

by Jason Earls

1
119
11199
1111999
111119999
11111199999
1111111999999
111111119999999
11111111199999999
111111111999999999
1111111111199999999999
1111111111119999999999
11111111111119999999999999
11111111111111199999999999999

Jackpot! That's exactly what we needed. (Notice we have an additional "1" over the total number of 9s in the number, which exactly matches Dr. Pickover's curio.) But what if our function z(n) could be simplified a little more to make it prettier? (We don't want to upset any "highbrow" mathematicians who might be reading this by listing "ugly" and "inefficient" functions.) So let's rewrite our function and change the name to h(n) for "highbrow" like this:

$$h(n) = (5^n * 2^{n+3} + 2^{2*n} * 5^{2*n} - 90)/90$$

Isn't that better? (Really, I was hoping it could be simplified even more since it still seems too long; can you find a more elegant simplification that produces the same numbers?) Testing it shows that h(n) produces the same numbers as z(n), so we're still cooking with grease here. After calling up a reverse function, I ran a simple program to test all the n values from 1 to 1000 to see if both h(n) and reverse(h(n)) were prime. But only h(7) turned up! Which is a major bummer because that's just the original value Dr. Pickover listed in his "tweet."

h(7) = 1111111999999

But at least we confirmed that his number and its reversal are both prime.

Since I couldn't find any larger values, let's take a moment to figure out what we're doing here exactly. Taking the idea down to its simplest form, basically we are searching for a subset of "emirps," which are primes such that their reversal is a different prime. Here are all the emirps up to 1000:

13, 17, 31, 37, 71, 73, 79, 97, 107, 113, 149, 157, 167, 179, 199, 311, 337, 347, 359, 389, 701, 709, 733, 739, 743, 751, 761, 769, 907, 937, 941, 953, 967, 971, 983, 991,

Does that sequence help us at all? Nope. Not that I can see. But it does give me an idea. How about we let r(n)=repunits (1, 11, 111, ...) and look for the least integer k such that both r(n)+k and its reversal are both prime numbers? Although these numbers won't follow the exact pattern exhibited by Dr. Pickover's original curio, a few may still be somewhat close. Running a computer search produces this data (values here are listed as "**n:k**" and will be explained further on):

2:2, 3:2, 4:40, 5:38, 6:8, 7:228, 8:6, 9:20, 10:102, 11:2, 12:38, 13:42, 14:146, 15:416, 16:42, 17:216, 18:20, 19:732, 20:702, 21:46, 22:810, 23:90,

24:1868, 25:346, 26:408, 27:658, 28:288, 29:308,
30:500, 31:508, 32:2868, 33:212, 34:18, 35:1326,
36:2290, 37:5500, 38:32, 39:20, 40:12366,
41:3636, 42:1238, 43:6652, 44:582, 45:9026,
46:1648, 47:7946, 48:2918, 49:178, 50:1526,
51:1906, 52:10312, 53:2450, 54:3208, 55:900,
56:498, 57:5968, 58:232, 59:842, 60:8048,
61:8626, 62:3822, 63:2680, 64:1002, 65:13110,
66:2048, 67:1716, 68:4458, 69:5716, 70:8196,
71:1346, 72:14558, 73:3108, 74:306, 75:572,
76:18990, 77:11298, 78:3790, 79:11610, 80:1070,
81:9566, 82:6816, 83:6792,

For example, when n=10 we have k=102 which means
that r(10) + 102 = 1111111213 is prime and so is its
reversal. Look at the k value for n=40 above: 12366! The
computer had to crunch quite a few numbers to find that
little beauty. (It seems the k values grow larger as n gets
larger, but might there be an upper bound?) This
sequence is pretty damn neat since it causes me to make
this conjecture:

**Conjecture: for any positive integer $n > 1$, there
will always be a positive integer k such that r(n) +
k and its reversal are both prime numbers, where
r(n)=repunits.**

We're still no closer to finding a larger version of
Pickover's original curio, but at least we procured a nice
conjecture that will probably take a thousand years or
more to prove by someone with a lot more brainpower
than I possess. You can't beat a deal like that.

Anyway, now we shall define another sequence using the
original h(n) function. We are going to see if any h(n)

terms and their reversals both have the same number of prime factors. Perhaps we can find a pattern or two there that will yield something interesting. Hunker down and examine this data for a spell:

h(3): 11199 : 99111 : 2
h(4): 1111999 : 9991111 : 3
h(7): 1111111999999 : 9999991111111 : 1

This means that the value for h(3) and its reversal both have two prime factors. While h(4) and its reversal both have three prime factors. And of course for h(7), it and its reversal have only one prime factor since they are both prime (Pickover's original curio). Now let's actually look at some of the prime factors to better understand the definition of this sequence. Here is the value for h(17) and its factors:

11111111111111111199999999999999999 =
157 * 70771408351026191082802547707

While its reversal and factors are:

99999999999999999911111111111111111 =
4381640223041 * 22822503653802187271

Only two for both. Are there more terms that match the definition of this sequence? Yes. Here are all the known *n* values such that h(*n*) and reverse(h(*n*)) both have the same number of prime factors:

1, 3, 4, 7, 17, 19, 20, 28, 41

That's all I could find so far. The difficulty of complete factorization of large numbers makes obtaining terms for this sequence pretty hard. Nevertheless, I will do this:

Conjecture: There are infinitely many n such that h(n) and its reversal both have the same number of prime factors!

Questions

1. Is it possible to simplify the h(n) function even more?

2. Can you prove any of the conjectures given in this article?

3. Would you like to extend any of the data given above?

379009*n and Digital Sums

"A work of art is above all an adventure of the mind."
-Eugene Ionesco

"'Curiouser and curiouser,' cried Alice."
-Lewis Carroll, Alice's Adventures in Wonderland

The number 379009 spells "Google" when turned upside down on a calculator. And 379009 is also a prime number (it possesses no divisors other than 1 and itself).

For some inexplicable reason, I recently got the idea to run a computer search to hopefully find numbers that had 379009 as one of their prime factors, yet the numbers also had to have very low digital sums (i. e. the sum of digits was a (relatively) small value – examples will be given later.)

Now why would I want to do such a thing? What purpose would it serve? Would it ever have any practical value in the "real world"? How did this bizarre idea ever pop into my brain in the first place? I have no clue. But once an idea comes to me, however weird it might be, I have no choice but to simply follow my whims and compulsions wherever they might lead since I have no control over myself when it comes to doing these mathematical activities. And I don't ask too many questions about whatever strange "creative" process I possess, if that's what it even is. I can't resist these strange numerical notions that come to me for whatever reason and I just go with my eerie brain-flow and hold on for the ride. Enough.

First, let's compute a few numbers of the form 379009*n along with their digital sums and see just how large those digital sums might be (for example, 379009*3 = 1137027 and 1+1+3+7+0+2+7 = 21). Here's some preliminary data in the form: "**n:digitsum(379009*n)**":

1:28, 2:29, 3:21, 4:22, 5:32, 6:24, 7:25, 8:17, 9:18, 10:28, 11:38, 12:30, 13:31, 14:23, 15:33, 16:25, 17:26, 18:27, 19:19, 20:29, 21:48, 22:40, 23:32, 24:33, 25:34, 26:35, 27:18, 28:19, 29:29, 30:21, 31:40, 32:32, 33:33, 34:34, 35:26, 36:27, 37:19, 38:20, 39:30, 40:22, 41:41, 42:42, 43:43, 44:44, 45:27, 46:28, 47:29, 48:30, 49:31, 50:32, 51:42, 52:43, 53:35, 54:36, 55:37, 56:20, 57:21, 58:31, 59:23, 60:24, 61:34, 62:44, 63:45, 64:37, 65:38, 66:30, 67:31, 68:32, 69:24, 70:25, 71:44, 72:45, 73:46, 74:38, 75:39, 76:40, 77:41, 78:33, 79:34, 80:17, 81:45, 82:37, 83:38, 84:39, 85:31, 86:41, 87:42, 88:34, 89:26, 90:18, 91:46, 92:47, 93:39, 94:40, 95:32, 96:42, 97:43, 98:35, 99:36, 100:28.

So scanning through this small sample of terms it appears the digital sums fall primarily in the 20s through 40s range. But what would happen if we just picked a number at random, say 13, and tried to determine if numbers of the form 379009*n can ever have digital sums less than that particular value? Let's run a quick program and find out.

Math Freak
by Jason Earls

Computing, programming, computing away...

Wow! Here are a bunch of values that have digital sums of 11 and 12 respectively (format below is "**n: digitsum(379009*n): 379009*n**":

26678: 11: 10111202102
266780: 11: 101112021020
316668: 12: 120020022012
528779: 11: 200412000011
2717669: 11: 1030021010021
3166680: 12: 1200200220120
5287790: 11: 2004120000110
5303568:12:2010100004112

Examining the individual digit values in the right-hand column, of course they are going to be low (such as 4, 3, 2, 1, or 0.) Notice the numbers for the first two "12" digital sum values have decimal expansions somewhat similar in form. More on that in a minute. But first, I wonder if there will ever be a number of the form 379009*n that has a digital sum lower than 11? Let's try to find one real quick.

Computing, programming, crunching away...

Yes! This value has a digital sum of 10! Look:

29049445: 10: 11010001100005

Can you believe it? And the form of the number is totally different from the previous "11" and "12" values. Wait a minute. Aren't we forgetting something? Isn't there supposed to be a factor of 379009 in there somewhere? Remember, that was one of the main criteria that popped into my head when I made up this problem. Of course we're still multiplying all the numbers by 379009, but let's just make sure that factor is *completely visible* inside the factorization:

factor(11010001100005) =
[5]
[29]
[200341]
[379009]

There she blows. It's the largest prime factor. Just making sure.

Now we shall turn our attention to various patterns in this class of numbers that ever-so-slightly **KILLS** the interest factor. First though, we need to run another computer search to see if we can find a number with a digital sum lower than 10.

```
Computing, programming, crunching tons of
numbers...
```

OH MY GOD! Here is one with a digital sum of only 9:

348276690: 9: 132000000000210

Notice with this number we actually found a pattern we can work with to give an example of what we *DO NOT WANT* among the "solutions." I'm not sure how I missed it, but actually there is a smaller number with a digital sum of 9 and a factor of 379009:

$$\text{factor}(13200000000021) =$$
$$[3\text{\textasciicircum}2]$$
$$[379009]$$
$$[3869741]$$

Notice it's the same number as above, but with one of its final zeros stripped off. This allows us to adequately demonstrate how we want to find numbers of the form 379009*n with low digital sums, but we also want them to have *unique factorizations* from all previous numbers. Look at this larger sample of numbers with digital sums of 11 or less:

```
   26678: 11: 10111202102
  266780: 11: 101112021020
  528779: 11: 200412000011
 2667800: 11: 1011120210200
 2717669: 11: 1030021010021
 5287790: 11: 2004120000110
 5809889: 11: 2202000220001
26678000: 11: 10111202102000
27176690: 11: 10300210100210
29049445: 10: 11010001100005
34827669: 9: 13200000000021
52877900: 11: 20041200001100
58098890: 11: 22020002200010
```

Again, notice the patterns in the right-hand column. A few of the numbers have the same basic "structure," if you will. Keep in mind, we are only interested in solutions that have completely different prime factorizations. That is, we don't want numbers with basically the same prime factors but then just add a bunch of 2s and 5s to increase the value by a factor of ten (which simply adds zeroes to the end). Anybody can do that. And it ruins the compelling mystery and supreme beauty of number theory. We only want solutions with unique prime factorizations that still have the lowest possible digital sums...

Questions

1. Will numbers of the form 379009*n ever have a digital sum *lower* than 9?

2. Can you think of an efficient algorithm to significantly speed up the search for this class of numbers?

3. Are there only a finite amount of 379009*n numbers with an *exact* digital sum of 9? Can you rigorously prove it?

4. I use the freely available *Pari/GP* algebra package for most of my computations, would *MATHEMATICA* or another programming environment be better for finding these types of numbers?

5. Will there ever be a practical purpose on earth for the problems proposed and suggested by this article?

6. Do you think your I.Q. will increase at all by reading and thinking about the subject of this paper – even by just a little bit, say 3 or 5 I.Q. points?

Math Freak
by Jason Earls

7. Does it feel like there's some sort of block or confining net surrounding your brain and if you could just bust through it you would be able to solve all the problems presented in this paper; or even all the number theory problems in the known universe?

Mersenne's Mistake

A single candle glowed in the cell of the monastery, its yellow rays reflecting off the copper-colored walls. A monk sat at his small writing table hunched close to the flame. A wooden bed with no mattress set in a far corner, a dusty crucifix suspended over the headrest. The monk dipped his quill in ink, crouched closer to the parchment and wrote the final sentences of his letter: "I am determined to drive them out. We must rid the land of their miasmic presence. There will be no sorcery here when I complete my studies."

He signed the letter "Marin Mersenne," folded it twice and slid it under his Bible. Then he crossed himself and took another sheet from the corner of his table and began writing again: "Dear Fermat, in the past week I have been investigating multi-perfect numbers. I recently made the discovery that..."

It would take Mersenne many hours to catch up on his correspondence. But after thirty minutes of writing, he paused to light another candle, then tapped his quill against his chin. He tugged at the collar of his robe, then bent to the parchment once again. Another two hours passed and finally he placed the letter with the others. He blew out the candle, stood and went to his bed in the darkness. Kneeling, he folded his hands to pray, and his whispered words filled the dark cell.

* * *

Math Freak
by Jason Earls

"But we must do something about the sorcery that is spreading," said Mersenne.

The Abbot paced behind his desk, hands clasped behind his back. Mersenne could see his knuckles turning white.

"Are there really that many sorcerers to worry about, Marin?"

"Yes. And the worst thing is they look to various magicians documented in the scriptures for inspiration. They extrapolate upon whatever esoteric biblical accounts they can find and distort the facts beyond reason."

The Abbot stopped pacing. He looked out his window to the fields behind the monastery. "Continue, Marin. Tell me more."

"Well... the sorcerers study apocalyptic scriptures and attempt to work their magic in the most vile ways. They are beginning to exert an influence over the common people. We must do everything we can to wipe out their brand of witchcraft as soon as possible."

The Abbot turned to Mersenne with his brow furrowed. "But shouldn't we focus on more immediate concerns? Such as feeding the poor. Inventing new methods of irrigation. Spreading the Word of God. Perhaps we shouldn't focus so much on the negative."

"Those things are very important, of course." He cleared

Math Freak
by Jason Earls

his throat. "But if we do not deal with the rampant
Satanism that's spreading, there will only be dire
consequences. The sorcerers are becoming too powerful,
too plentiful. Lately I have spent much time studying
various ways to invalidate their magic and to convince the
common people the sorcerer's actions are the devil's
work."

The Abbot sat down. He leaned forward and propped his
elbows on his desk. "I want you to continue your scientific
work, Marin. But I will no longer permit you to study
ways to fight sorcery or witchcraft. You must quit this
obsession with Satanism. It isn't healthy."

"But..."

"No. I will not provide any reasons for my decision
because I won't consider any of your protests."

Marin stared into the Abbot's eyes. He looked into them
for a long time. He did not understand the decision, but
knew he must obey. He stood and retied the black cord
girding his robe, then looked at the Abbot again. He knew
the decision was final so he quickly turned and left.

* * *

Walking through the countryside, Mersenne stared at the
hundreds of rabbits that lay slaughtered in the fields.
"Evil," he whispered, then lowered his head and tucked
his bundle of letters closer to his side.

Math Freak
by Jason Earls

Some of the rabbits were still alive. They squealed and kicked up dust in the fields and Mersenne covered his ears and looked away. But then his eyes went back to the rabbits. Row after row of them. Mostly dead but a few still shaking and kicking.

He passed the fields and came to three small hills covered with tall grass. Large circles were burnt into the hillsides. Black rings with grey smoke rising from the centers. He crossed himself and recited, "The Lord is my shepherd; I shall not want. He maketh me to lie down in green pastures ..." Then he quickened his step down the dirt road.

He came to six wooden crosses that he'd never noticed before. They came up to his shoulders and were partially burnt. Drops of blood were dripping down the sides. Bright red liquid spreading down the black, forming puddles on the ground.

"Where is all this evil coming from?" he said.

He shut his eyes tightly, squeezed his bundle closer and began to run.

* * *

Two pots of ink set at the corner of Mersenne's table. A pile of manuscripts and a stack of blank parchment at the other. The table teetered with thin strips of wood tucked under the legs. He was surprised it had stayed upright for so many years. He looked at the pattern in the wood grain, thought for a moment, then glanced to the crucifix above his bed. He smiled, bent forward and wrote, "Concerning numbers of the form $N = 2^p - 1$, p must be prime for N to be prime. This is easy to show with elementary algebra. But I will make this claim: For N to be prime, these values of p are necessary: 2, 3, 5, 7, 13, 17, 19, 31, 67, 127, 257, with no others less than 257."

Mersenne was writing to Frenicle de Bessy on the material that would be in his next book. For many years Mersenne had searched for a formula that would produce every consecutive prime number and now he felt he was on the verge of a breakthrough. While working on the formula, he had arrived at a conjecture for the sequence of p values that make $2^p - 1$ prime. He planned to include the material in the fourth chapter of his book.

Even though he could not test the values directly, Mersenne felt fairly confident the sequence was correct. He and Frenicle had discussed a primality testing method for several months and had developed a secret algorithm that they had not revealed to their other correspondents. They both had doubts about the proof underlying the process however -- they suspected it wasn't fully rigorous.

Mersenne wrote another paragraph, then finished with the line, "I hope one of us will soon find the missing step to prove the foundation of our algorithm. I think the formula for producing every consecutive prime is well within our grasp. But even if we do not locate the missing step, surely a formula that produces only prime numbers (even if they are not consecutive) will soon be attainable."

Math Freak
by Jason Earls

He signed the letter, folded it twice and placed it at a far corner. He rose from the table, went to his bed and lifted a Bible from his pillow, along with a copy of *Arithmetica* by Diophantus which lay beside it. Then Mersenne turned and quickly shuffled out of his cell.

* * *

A crowd of monks were gathered at the large tree behind the monastery. Mersenne stood watching them from his window. A few were standing and crossing themselves while others bowed and glanced up into the branches of the tree. They were pointing at something. Mersenne walked through the monastery, went out to the group and looked up at the tree.

The Abbot was swaying there among the twisted branches. Suspended by a thick rope encircling his neck. Dead rabbits were hanging next to him, with frayed twine around their midsections. The Abbot's limbs were broken, bent backward against the joints unnaturally. Strings of meat were dangling from his throat. Mersenne counted six holes punctured in his lower abdomen, his intestines pulled out and hanging. The bloody cords swayed next to the dead rabbits.

A monk next to Mersenne began to weep while others pointed and talked in low voices. Marin looked at the Abbot again and noticed his feet were missing. Hacked off at the ankles. Mersenne searched the ground but could not see them. He lowered his head and crossed himself, then slowly walked back to his cell.

Once there, he got down on the floor and lay on his stomach. He began reciting Matthew 10:1 – "And when he had called unto him his twelve disciples, he gave them power against unclean spirits, to cast them out, and to heal all manner of sickness and all manner of disease."

He quoted other scriptures and mixed them with short prayers, keeping his eyes shut and pressing his palms against the cool floor.

Then he felt a presence in the room.

A strong presence of evil.

Mersenne opened his eyes and felt a force pulling his attention upward. He tried to resist, but the presence was too powerful. His head started to rise involuntarily and he strained to pull it down. But he was forced to yield to the presence. He pushed himself up, leaned back on his knees and blinked until his vision adjusted. He saw a small figure hovering in the air next to his crucifix.

A tiny man. Elvish, yet possessing evil-looking facial features. Slanted yellow eyes, a long crooked nose. He floated and swayed in the air, dressed in a black robe with green designs. A silver necklace with symbols Mersenne had never seen before dangled from his neck. He was cradling something in his thin arms. Mersenne squinted, trying to see what it was.

Math Freak
by Jason Earls

A creature. Definitely not human.

The small man held it with care as if it were his child. Mersenne got to his feet and stared into the elvish man's face, then looked at the creature. It seemed to have too many limbs and no eyes. It squirmed and moved its mouth unnaturally without making a sound.

"Marin Mersenne," said the hovering man.

Mersenne nodded.

"You are trying to rid the land of sorcery."

"Yes, I am." His voice cracked a little, he cleared his throat.

"You will fail." The little man grinned and swayed. "Our sorcery is too powerful and it has spread too far."

Marin crossed himself and looked at the creature squirming in the man's arms. He tried to think of a response. Sweat dripped down his back and seconds passed. Finally he was able to force away his fear: "No. God will help me stop your sorcery. My studies will soon be able to drive your kind from the Earth."

The sorcerer hovered, his yellow eyes went to the crucifix on the wall. He opened his mouth to speak but Mersenne interrupted him –

"Did you kill the Abbot?"

Math Freak
by Jason Earls

The man laughed, then changed subjects: "You think you're a scientist don't you, Mersenne?"

Marin narrowed his eyes.

"Do you think you are a scientist, or not? Answer me."

"I am a monk."

"But in your heart you greatly desire to be a scientist."

"You have no idea what is in my heart."

"My sorcery tells me many things. Yesterday you wrote a letter to Frenicle de Bessy concerning prime numbers. Your letter contained many mistakes. In your sequence of values for primes of the form 2^p-1, you made the claim that $2^{67}-1$ and $2^{257}-1$ are prime. But they are composite. You set forth a foolish hypothesis that your superficial human knowledge has no way of ascertaining."

"Perhaps I did. But your criticism does not bother me."

"Surely you don't think mathematics is exempt from my sorcery? Allow me to demonstrate my expertise."

Mersenne's eyes widened. He wondered what the sorcerer would say next.

The tiny man bent toward the creature in his arms. He leaned down and whispered into its ear. It released a soft

Math Freak
by Jason Earls

whine and raised three twisted arms to the sorcerer's face. Mersenne counted eight fingers on each hand.

"I will construct a prime from your mistakes, Mersenne. Keep in mind your non-prime values of $2^{67}-1$ and $2^{257}-1$."

The sorcerer shut his eyes. He placed his fingertips to his eyelids and began to sway. Mersenne didn't blink.

"If you concatenate the values of $2^{67}-1$ and $2^{257}-1$ with 68 zeros between them, you will get a prime with the decimal expansion of

147573952589676412927000000000000000000
00000000000000000000000000000000000000
0000000000000002315841784746323908471419
70017375815706539969331281128078915168015
826259279871."

The sorcerer recited each digit aloud.

Mersenne wiped sweat from his chin. "Are you positive that is accurate? Is that really a prime number?"

"Of course it is. I can give you more if you like."

"Yes! I mean..." Mersenne tried to force away his intense curiosity. He lowered his head.

The sorcerer shut his eyes and floated in a wider arc. He concentrated so hard he rose to the center of Mersenne's cell. The creature in his arms squirmed and moved its mouth awkwardly. Mersenne noticed thin insect-like wings folded under its back.

"I see the pattern in the numbers now," said the sorcerer. "It is complicated and buried deep within their natural order, but I can see the correct values that make them prime. Let $2^{67}-1$ be r and $2^{257}-1$ be s. The values of n that make $r * 10^n + s$ prime are 42, 62, 146, 210, 936, and 1490, with no others up to 4000. It means that with values greater than 62 – (because the first two values do not yield the desired decimal expansions) – if you concatenate the r and s values while inserting 68, 132, 858, and 1412 zeros between them, you'll get prime numbers."

Mersenne stepped backward and shook his head. He went to his desk and dipped his quill and wrote down the sorcerer's values with a note of explanation beneath. He turned and faced the floating man and suddenly felt anger burn through his body.

"Why are you here? What do you want from me!" Mersenned shouted.

"*Ha ha ha*. I want my magic to flourish, of course. I want a pestilence to spread throughout the land. I want to dispose of all Christians. I want demons and magicians and Satan himself to freely roam the Earth."

"Never. I will do everything within my power to prevent that from happening."

Math Freak
by Jason Earls

The creature moved its many appendages and the sorcerer laughed again. "You actually think you possess the power to stop me?" His eyes brightened.

"Yes. I do."

The magician stared at Mersenne. He looked beyond his eyes and into his mind and saw that Marin did have sufficient power to at least hinder his efforts.

"Unlimited mathematical knowledge," the sorcerer said.

"What?"

"I will give you unlimited mathematical knowledge."

He glanced to his writing desk. "I would never accept anything from you."

"And in return you will end your anti-witchcraft studies."

"Nothing you could ever give me would stop me from trying to annihilate you and your kind."

The sorcerer raised one hand and extended a finger, signaling Mersenne to wait. He bent toward the deformed creature and whispered into its ear again and Mersenne saw both of their eyes begin to glow green.

The sorcerer chanted a phrase in an unknown language and Mersenne's arms began to shake. A burning sensation passed throughout his entire body. His legs felt

Math Freak
by Jason Earls

heavy, he couldn't move. A dark presence traveled up his spinal column, and he felt a change occur inside him.

"You now have the gift," said the sorcerer.

"What gift. I told you I wanted nothing from you."

"You don't have a choice. It is done. You will notice the change the next time you write."

"No, you can't force any change upon me. I am a man of God."

"It doesn't matter. Evil is stronger than good."

"Never. You are wrong."

"It doesn't matter. You have the gift now. You must cease your anti-witchcraft studies. We will leave your land and practice in another part of the world and you will never try to stop us again."

Mersenne started to respond, but knew that arguing further was futile. He turned his back to the sorcerer, dropped to one knee and began to pray.

In the background he could hear the creature choking and gasping. The sorcerer whispered: "What shall we do now?"

"R-repeat the d-digits of the f-first prime to b-bring him d-down," the creature said.

Math Freak
by Jason Earls

Mersenne stayed in his kneeling position and listened to the sorcerer recite every digit of the initial prime's decimal expansion once again. He tried to ignore the digits and continue praying, but his appeal to God was not strong enough to prevent the sorcerer's magic from working. The sounds of the digits forced his hands to drop, his knees to give way, and he collapsed to the floor.

The sorcerer floated over to Mersenne's writing table where he rifled through his manuscripts. He took the Bible and letters and tucked them under his arm. He recited more words in an ancient language, then floated out the window with the creature.

* * *

The next morning Mersenne lay on his wooden bed. He couldn't remember what had happened. He stared at the ceiling, feeling the hard wood beneath him. Images of the sorcerer and the creature slowly returned. He tried to determine if the encounter had been a dream or not. He leaned up, glanced around his cell, noticed his Bible and letters missing. *Why would they want them?* He stood and retied his robe, then went to the dining hall for breakfast, never mentioning the sorcerer or the creature to his fellow monks.

An hour later, back in his cell, he took up his quill and began a new letter. "Dear Frenicle, last night I had a strange dream. I don't fully understand what happened, but I think a sorcerer and his..."

Math Freak
by Jason Earls

Then his hand began to spasm.

The quill dropped.

His fingers cramped and froze into painful positions. His hand lurched for the quill, grasped it, and he furiously began scribbling out reams of formulae for several large numbers. Gargantuan integers, the numbers consisting of at least a million digits. He felt no control over his hand as it wrote extravagant claims beneath each formula. "This is the first odd perfect number. Difficult to find because it has exactly 2,463,918 digits ..."

Other formulae and mathematical claims followed, even more astonishing than the previous. His hand listed sophisticated algorithms for factoring large integers and more definitions for strange numbers followed with full descriptions of their properties. Mersenne did not recognize or understand all the claims. A few things he remembered hearing from Frenicle de Bessy a few years ago, which at the time he did not fully comprehend.

Mersenne forgot about his letter to Fermat. His hand scribbled on and his mind tried to absorb the information. Numerical patterns and algebraic conundrums flowed and his body seemed detached from his hand's actions. Then he fully realized that the encounter with the sorcerer had not been a dream. His hand stopped writing and he stared at the formulae and algorithms and bizarre claims listed before him.

Math Freak
by Jason Earls

They slowly began to make sense. He could see their structure. See the logic underlying each one. The abyss of pure number theory was revealing itself to him with perfect clarity. And he became frightened, then exhilarated. The cold logic, the harrowing certainty, the absolute truth of pure mathematical form. The profound structure humans were not supposed to observe.

He shook his head, tossed down the quill and stood up. His hand lurched for it again. He grabbed the hand with his other one and stepped toward the window. *I will cut it off*, he thought. *I will slice it off at the wrist.*

Then his body moved back to the desk and a force pulled him down into the chair. He took up the quill again and put it to the parchment and wrote, "Formula for Primes," then a long string of symbols and instructions beneath.

He immediatcly recognized that the formula would produce each prime in consecutive order.

It was the formula he had spent his whole life searching for.

He investigated the formula and it only required fifteen minutes to determine its accuracy. He managed to write out the first 10 primes with over 50 digits each.

After four hours of mathematical work, Mersenne's hand stopped writing. He regained control and began another letter to Fermat. He wanted to write of all the mathematical claims and theorems he had seen, but

Math Freak
by Jason Earls

mainly he wanted to relate the formula for primes that the sorcerer had given him. To tell him how simple and elegant it was. But he set down the quill after a few minutes, realizing he would not be able to reveal the knowledge to anyone. He couldn't let them know since the gift came from an evil source. He would lose their respect. They would realize the knowledge was the devil's work.

Mersenne pondered how he might be able to reveal the prime formula with subtle hints. *Perhaps I can work it out from a different angle*, he thought. But after two hours of trying, he realized it was impossible.

Finally Mersenne lay down his quill and went to the window of his cell. Images of formulae and numerical patterns still lingered in his mind. He looked out at the fields for awhile, then leaned his head against the glass, shut his eyes and prayed. When he looked up again, he saw the creature the sorcerer had held during their encounter, hovering ten feet from his window. Its thin insect-like wings were fluttering and its mouth opening and closing in grotesque shapes. It still had no eyes. The creature raised its three arms with eight fingers, holding them out and reaching for Mersenne. But Marin covered his face and turned away. A few minutes later he looked up and the creature was gone.

* * *

Mersenne stopped writing to his scientific correspondents after receiving the sorcerer's gift. He

Math Freak
by Jason Earls

spent all his time investigating the formula for primes
and other mathematical enigmas. Each time he
would pick up his quill, his hand would produce yet
another deep mathematical insight or theorem that
would arouse his curiosity and compel him to investigate.
Many of his former correspondents wrote to inquire
about his sudden end to their discussions, but he never
responded, even though he wanted to share his new
knowledge with them.

And Marin Mersenne was the only man to have ever
known the formula for producing consecutive prime
numbers. But in 1648, it died with him.

On Brocard's Factorial Square Problem

"Mathematics reveals its secrets only to those who approach it with pure love, for its own beauty."
-Archimedes

"When I have clarified and exhausted a subject, then I turn away from it, in order to go into darkness again."
-Carl Friedrich Gauss

"I spent forty years looking for patterns in Pi, I found nothing."
-Sol, from the movie 'Pi'.

To begin this chapter, here's a quote directly from the MacTutor biography of mathematician Henri Brocard: "In 1876, Brocard asked if the only solutions to the equation $n! + 1 = m^2$, in positive integers (n, m), are (4, 5), (5, 11), (7, 71). This problem remains open. It is not even known whether there are only finitely many solutions."

To illustrate the above problem a little more explicitly, here are the only known solutions to finding $n! + 1$ values such that they are squares:

$$4! + 1 = 25 = 5^2$$
$$5! + 1 = 121 = 11^2$$
$$7! + 1 = 5041 = 71^2$$

I'm sure a ridiculous amount of computation has been

Math Freak
by Jason Earls

applied to this problem over the years, yet no one has ever found another solution. The purpose of this chapter is to look at the problem from a different perspective, computationally speaking.

First, let's compute the sequence of least k such that $n!$ + k is a square and see if any values come relatively close to being a solution. Here is how the aforementioned sequence begins, starting with $n=1$:

0, 2, 3, 1, 1, 9, 1, 81, 729, 225, 324, 39169, 82944, 176400, 215296, 3444736, 26167684, 114349225, 255004929, 1158920361, 11638526761, 42128246889, 191052974116, 97216010329, 2430400258225, 1553580508516, 4666092737476, 565986718738441, ...

See those three little "1" values early on in the sequence? Obviously those are the solutions Brocard himself discovered and that everyone's been talking about ever since. But after those three appear, notice the terms grow larger and larger very quickly. From the looks of this small amount of data, it doesn't seem like there will be any other small k values coming anytime soon; meaning the values only get further and further away from being squares.

Now let's alter our approach once again. What would happen if we computed the **number of decimal digits** of the k values in the sequence above to see if they continue to grow larger (which is known as a "monotonically increasing sequence")? After acquiring

Math Freak
by Jason Earls

some super-efficient PARI code courtesy of M. F. Hasler from the entry for the sequence above in the *On-Line Encyclopedia of Integer Sequences*, I was able to compute the first 300 terms of our new "number of digits of *k*" sequence. Here they are:

1, 1, 1, 1, 1, 1, 1, 2, 3, 3, 3, 5, 5, 6, 6, 7, 8, 9, 9, 10, 11, 11, 12, 11, 13, 13, 13, 15, 16, 16, 18, 18, 18, 20, 21, 21, 22, 23, 24, 25, 25, 25, 27, 28, 29, 29, 30, 31, 32, 33, 34, 35, 35, 36, 37, 38, 39, 40, 40, 40, 42, 43, 44, 45, 46, 47, 48, 49, 50, 50, 52, 52, 53, 54, 55, 56, 57, 58, 59, 60, 61, 62, 62, 63, 65, 66, 67, 68, 68, 70, 71, 72, 73, 73, 74, 75, 77, 77, 79, 80, 81, 81, 82, 83, 85, 86, 86, 88, 88, 90, 91, 92, 93, 94, 95, 96, 95, 97, 99, 100, 101, 102, 102, 104, 105, 106, 107, 108, 109, 109, 111, 113, 114, 115, 115, 117, 118, 119, 120, 121, 121, 122, 125, 125, 127, 128, 129, 130, 131, 130, 133, 134, 135, 136, 137, 138, 139, 140, 142, 143, 144, 145, 145, 147, 148, 149, 151, 152, 153, 154, 155, 156, 157, 158, 160, 160, 161, 163, 164, 165, 166, 167, 169, 170, 171, 172, 172, 174, 175, 176, 177, 179, 180, 180, 182, 183, 184, 185, 187, 188, 188, 190, 191, 192, 194, 195, 196, 196, 198, 199, 201, 201, 203, 204, 205, 207, 208, 209, 210, 210, 212, 213, 215, 213, 216, 218, 220, 220, 222, 223, 224, 225, 227, 228, 229, 228, 231, 233, 233, 235, 236, 237, 238, 239, 241, 242, 243, 244, 245, 247, 248, 249, 250, 251, 252, 254, 255, 257, 257, 259, 260, 260, 262, 262, 265, 266, 267, 269, 270, 270, 272, 273, 275, 276, 277, 278, 280, 281, 281, 283, 284, 285, 287, 288, 289, 290, 290, 292, 294, 295, 297, 298, 299, 301, 301, 303, 304, 305, 306, 307, ...

Math Freak
by Jason Earls

Study the data above carefully. Notice there are a few instances where the sequence *does not* strictly increase: (**11, 12, 11, 13**) and (**95, 96, 95, 97**) and (**129, 130, 131, 130**) for example, but for the most part the terms do continue to grow larger overall. After computing the first 15,000 terms of the sequence above, I know for certain that for all integers $n > 19$ up to that limit (15,000), there is not a value in the sequence with less than **10** decimal digits. I don't know about you, but that pretty much convinces me that another "1" is not going to be appearing anytime soon. But of course in the unpredictable world of numbers and computational number theory that doesn't mean jack-squat because anything can happen if you don't have a rigorous proof. Nevertheless I am still going to make this conjecture:

Conjecture: In the sequence of "least k such that $n! + k$ is a square," for all $n > 7$, the k value will never have less than 2 decimal digits.

Shouldn't a mathematician somewhere be able to prove there will never be anymore square $n! + 1$ values based on the ideas we have outlined in this note? I mean, with this kind of computational evidence now visible, doesn't it convince you there won't be any more "1" solutions coming? So why can't a mathematician somewhere just go ahead and prove this already? Come on, what the **hell** is wrong with these **damn people**? Isn't there some type of function underlying our idea that could be found and then rigorously proved to grow bigger and bigger continually and thus this old problem could be settled once and for all? What is the deal with these *bloody*

Math Freak
by Jason Earls

mathematicians anyway? Why can't they prove anything good nowadays? What in the world are they getting paid to do? Blow hot air and teach kids how to solve train departure problems? Maybe they should just all burn in... ***OF COURSE THIS RANT IS ONLY A JOKE. I AM ONLY KIDDING. MATHEMATICIANS ARE AWESOME.*** Real mathematicians are incredible and vastly superior to myself since I will never be **smart** enough to be one.

Seriously though, if someone could prove our sequence is "*roughly*" monotonically increasing, (meaning the terms may backpedal slightly every once in a while, but for the most part they will continue to grow), then this entire Brocard factorial square problem could be resolved forever. That's all there would be to fully proving it, which sounds pretty easy, right? But filling in the exact details for a fully rigorous proof will still be a bitch.

Computations on the Reverse-and-Add 196-Problem

"A thorough investigation of the concept of number must always turn out to be somewhat philosophical."
-Gottlob Frege

"Most people would sooner die than think; in fact, they do so."
-Bertrand Russell

"If someone believes in mathematical objects and their queer properties--can't he nevertheless do mathematics? Or--isn't he also doing mathematics?"
-Ludwig Wittgenstein

Here's a catchy little algorithm that gives rise to a deep problem in elementary number theory:

1. Pick a number.
2. Reverse its digits and add the result to the original number.
3. Repeat the process until you arrive at a palindrome (a number that reads the same way forward and backward).

Let's have an example. We'll choose 49. 49 + 94 = 143, and 143 + 341 = 484, a palindrome! But not every number we might select eventually reaches a palindrome. The integer **196** is the first cantankerous number refusing to produce a palindrome under this simple "reverse-and-add" rule, even though it has been added and reversed until the result has literally millions of decimal digits!

Most numbers however do reach a palindrome after only a small amount of steps (iterations). Here is data showing the exact number of iterations necessary for every number from 1 to 100 to reach a palindrome *(format below is "number:steps")*:

1:1, 2:1, 3:1, 4:1, 5:1, 6:1, 7:1, 8:1, 9:1, 10:2, 11:1, 12:2, 13:2, 14:2, 15:2, 16:2, 17:2, 18:2, 19:3, 20:2, 21:2, 22:1, 23:2, 24:2, 25:2, 26:2, 27:2, 28:3, 29:2, 30:2, 31:2, 32:2, 33:1, 34:2, 35:2, 36:2, 37:3, 38:2, 39:3, 40:2, 41:2, 42:2, 43:2, 44:1, 45:2, 46:3, 47:2, 48:3, 49:3, 50:2, 51:2, 52:2, 53:2, 54:2, 55:1, 56:2, 57:3, 58:3, 59:4, 60:2, 61:2, 62:2, 63:2, 64:3, 65:2, 66:1, 67:3, 68:4, 69:5, 70:2, 71:2, 72:2, 73:3, 74:2, 75:3, 76:3, 77:1, 78:5, 79:7, 80:2, 81:2, 82:3, 83:2, 84:3, 85:3, 86:4, 87:5, 88:1, 89:25, 90:2, 91:3, 92:2, 93:3, 94:3, 95:4, 96:5, 97:7, 98:25, 99:1, 100:2.

Notice that most of the terms above resolve in less than a few iterations, but 89 takes 25 steps! Here is the full "trajectory" for 89:

89-> 187-> 968-> 1837-> 9218-> 17347-> 91718->173437-> 907808-> 1716517->
8872688-> 17735476-> 85189247-> 159487405->
664272356->1317544822->
3602001953-> 7193004016-> 13297007933->

Math Freak

by Jason Earls

47267087164-> 93445163438 -> 176881317877->
955594506548-> 1801200002107->
8813200023188.

Look at the nice palindrome at the very end of the "cycle."
It's rather surprising it takes so many iterations to arrive
at that number when its neighbors resolve so much
faster. Concerning numbers that no one on Earth knows
if they'll ever reach a palindrome, they are known as
"Lychrel" numbers and the sequence begins like this: 196,
295, 394, 493, 592, 689, 691, 788, 790, 879, 887, 978,
986 ...

Quite a few of them. Although you should be aware that
some of those terms fall into trajectories of other Lychrel
numbers, i. e. they join in on the same path that others
take after a few iterations. The Lychrels that are not
known to join any other paths are considered "pure" and
called "seeds."

Lychrel numbers were named by Wade VanLandingham
based on the letters of his girlfriend's name (Cheryl).
VanLandingham runs an excellent web site (p196.org),
devoted solely to the 196-algorithm and Lychrel numbers,
where he keeps track of current world records and
provides related news items. VanLandingham presently
holds the world record for the 196 palindrome quest and
has computed its trajectory for nearly 725 million
iterations with a resulting number of over 300 million
digits! Yet he still has not found a palindrome.

Concerning 196, let's examine its actual trajectory and see

if we notice anything unusual. Here is the beginning of the reverse-and-add algorithm starting with 196:

196-> 887-> 1675-> 7436-> 13783-> 52514->94039-> 187088-> 1067869->10755470-> 18211171-> 35322452-> 60744805->111589511->227574622->454050344-> 897100798-> 1794102596->8746117567-> 16403234045->70446464506-> 130992928913-> 450822227944-> 900544455998->1800098901007-> 8801197801088.

Looking at the terms carefully, it's intriguing that a couple of them are very close to being palindromic. Notice the term 897100798 for example. It's just one digit "off" in its decimal expansion: reading left to right, if the second '0' would have been a '1' we would have had *897101798* and the iteration process would have halted right there. Also observe the last term in the trajectory, 8801197801088. It's only two digits off from being palindromic: reading left to right, if the '80' after the first 7 would have been a '91' it would have made the palindrome:

8801197911088

So close, yet so very far away, as the cliché goes. As you can see, only one or two little digits can make all the difference in the world when working with palindromes!

Miraculously, people have been reversing and adding on the 196 starting number for many years now (Van

Math Freak
by Jason Earls

Landingham halted his research in 2006 after arriving at a 300 million digit number) and today no one knows whether it will eventually land on a palindrome or not. Imagine, such a simple process, yet some of the most brilliant mathematicians on Earth have never proved that 196 under the reverse-and-add algorithm will, or will not, reach a palindrome, even though a ton of computational evidence points in the latter direction. It brings to mind this quote by W. S. Anglin:

> *"Mathematics is not a careful march down a well-cleared highway, but a journey into a strange wilderness, where the explorers often get lost. Rigor should be a signal to the historian that the maps have been made, and the real explorers have gone elsewhere."*

Another interesting trajectory is the one for the (conjectured) Lychrel number 879. Let's look at its terms and see if any are close to being palindromic:

879-> 1857-> 9438-> 17787-> 96558-> 182127->903408-> 1707717-> 8884788->17759676-> 85455447-> 159910905-> 668930856-> 1326970722-> 3597766953-> 7194444906-> 13288889823-> 46187778054-> 91275556218->172541113437.

Notice above we also have two terms that come very close to being palindromes. Consider the term 8884788, for example. Reading left to right, if the 3rd '8' would have been a '7' it would be a palindrome. Also look at the 3597766953 term. Reading left to right again, notice that

Math Freak
by Jason Earls

if '66' were changed to '77,' or the '77' changed to '66,' it would be a palindrome. So there are two terms early on in the 879 trajectory that come close to halting the reverse-and-add algorithm, but again no cigar!

I wonder if all Lychrel numbers are stubborn in this sense: that they glide along, almost becoming palindromic by just one or two digits being off, but then never actually become one?

What about numbers that take a longer amount of "time" than others to eventually reach a palindrome in the 196-problem? Out of all integers less than 10 million, the term 9008299 requires 96 iterations to reach a palindrome. Let's round 96 up to a 100, then split it in half to get 50 and use that as a benchmark so we can have fun engaging in a little computation of our own. That is, we will attempt to find terms that require at least 50 iterations to become palindromes under the reverse-and-add algorithm, which might perhaps better aid us in understanding the behavior of Lychrel numbers. After writing a quick and dirty program and running it on a spare ancient computer down in my basement, the following sequence of 21 terms popped up that each take at least 50 iterations to reach a palindrome. The exact number of iterations is listed after each term *(format again is "number:steps").*

10677:54, 10833:55, 10911:56, 11667:54, 11823:55, 11901:56, 12657:54, 12813:55, 13647:54, 13803:55, 14637:54, 15627:54, 16617:54, 17607:54, 20676:54, 20832:55, 20910:56, 21666:54,

21822:55, 21900:56, 22656:54,

But unfortunately these terms don't tell us much about the behavior of Lychrel numbers, since (sadly) all of their trajectories ultimately lead to this mysterious palindrome:

4668731596684224866951378664

As for performing more computations of our own, next we will try to test one of the original Lychrel "seeds" and see if we can make it reach a palindrome. Judging from Wade VanLandingham's web site, it appears the seed number 10583 has not been "crunched" very much. So it
. will be our choice for a few thousand iterations.

Alas, 10583 seems to be another stubborn bird. It took a little over one hour for a program to iterate it exactly 25000 times, yet it never reached a palindrome.

Could any Lychrel numbers ever be consecutive? That is, can two numbers n and $n+1$, (neighbors on the number line), both not be known to reach a palindrome after quite a few (say 150) iterations? Yes. Here is the sequence of n and $n+1$ such that both are (conjectured) Lychrel numbers (they didn't resolve after 150 iterations):

**8989:8990, 10787:10788, 11777:11778,
12767:12768, 13392:13393, 13596:13597,
13757:13758, 14096:14097, 14382:14383,
14586:14587, 14747:14748, 15086:15087,
15372:15373, 15576:15577, 15737:15738,**

Math Freak
by Jason Earls

**16076:16077, 16362:16363, 16566:16567,
16727:16728, 17066:17067, 17352:17353,
17556:17557, 17717:17718, 18056:18057,
18096:18097, 18342:18343, 18546:18547, ...**

My program continued spitting them out quickly and
easily, which seems to be adequate computational
evidence that these consecutive Lychrel numbers are
infinite. But since no one has yet succeeded in proving
that 196 (or any other "seed" candidate) are indeed true
Lychrel numbers, the conjecture that there are infinitely
many consecutive Lychrels seems light years off from
having a chance at being proved either true or false!

In summary, hopefully this chapter has provided a few
"slant views" on the 196-problem and Lychrel numbers in
general. Perhaps we were able to shed a little light on how
Lychrel numbers and their iterative process functions. Or
if we have just encouraged one brave soul to further
investigate this captivating problem, or even allowed
them to understand the overall details better, this chapter
has done its job. Maybe an interested reader would like to
extend one or two of the computations given above, or
perhaps try to take the problem in a completely different
direction. One of the great things about mathematics is
that it's highly flexible and ripe for intense exploration.
So boot up a spare ghetto computer that you're not using
and let your fingers rip dexterously across the keyboard
as you crunch some ginormous numbers and perform a
few million iterations of your own. But be sure to guide
everything with your wise mind. And to have as much fun
as possible.

Brilliant Harshad Numbers are Finite

"Infinity is unobtainable by the mere brute force
of computerized number crunching."
-Simon Singh

"I and this mystery, here we stand."
— Walt Whitman, Song of Myself

A Harshad number is one that is divisible by the sum of its own digits. For example, 1236 has a digital sum of 1+2+3+6 = 12 and 1236/12 = 103, so it's a Harshad number. A brilliant number is one that has exactly two prime factors of the same length (using base 10). For example, 3007 has just two prime factors of two digits each: 3007 = 97*31, so it's brilliant.

Could there be infinitely many brilliant Harshad numbers? Nope, not a chance. And I can prove they are finite (thanks to Dr. Bart Snapp who sent me an outline of a proof along with two main functions that you'll soon see below; and also to Dean Hickerson who first sent me a proof many years ago but I lost it and can't remember how it went). First, let's run a computer search and see if we can find at least a few brilliant Harshad numbers. Some lines of Pari code quickly reveal this sequence to be all the brilliant Harshad numbers anyone will ever find:

4, 6, 9, 10, 21, 209, 247, 407, 481,
629, 803, 1387, 1679.

Math Freak
by Jason Earls

Only 13 of them. It sucks there will never be any more, but we'll just have to face up to the harsh fact that 1679 is the last brilliant Harshad number anyone will ever find floating throughout the Platonic Realm (if you believe in such things).

To prove brilliant Harshads are indeed finite, first we need a function representing the maximum sum of digits a number can have. We'll call it "maxsod" for maximum sum of digits, and define it thus:

$$\text{maxsod}(n) = 9*\text{floor}(\log 10(n)+1)$$

Next, we need a function representing the smallest the prime factor for a brilliant number n can possibly be and we shall call it "minspf" for minimum smallest prime factor, and define it like so:

$$\text{minspf}(n) = 10\char`\^(\text{floor}(\log 10(\text{sqrt}(n))))$$

Basically our proof is going to involve giving evidence that when n is a brilliant number and sufficiently large, then **maxsod(n) < minspf(n)**, and therefore the sum of digits will be too small to ever be a factor again.

Before doing that, we first need to ask a simple question that will help us find a permissible range to search through to make sure no other brilliant Harshads exist. After finding this range and searching through it, we can then be quite confident we have found them all. To find this permissible range, we can simply compare the two

Math Freak
by Jason Earls

functions above and see when for brilliant numbers n
maxsod(n) >= minspf(n), which means the maximum
sum of digits is greater than or equal to the minimum
smallest prime factor.

This is the full sequence produced for the permissible
range: 4, 6, 9, 10, 14, 15, 21, 25, 35, 49, 121, 143, 169, 187,
209, 221, 247, 253, 289, 299, 319, 323, 341, 361, 377, 391,
403, 407, 437, 451, 473, 481, 493, 517, 527, 529, 533, 551,
559, 583, 589, 611, 629, 649, 667, 671, 689, 697, 703, 713,
731, 737, 767, 779, 781, 793, 799, 803, 817, 841, 851, 869,
871, 893, 899, 901, 913, 923, 943, 949, 961, 979, 989,
1003, 1007, 1027, 1037, 1067, 1073, 1079, 1081, 1121,
1139, 1147, 1157, 1159, 1189, 1207, 1219, 1241, 1247, 1261,
1271, 1273, 1333, 1343, 1349, 1357, 1363, 1369, 1387,
1403, 1411, 1457, 1501, 1513, 1517, 1537, 1541, 1577, 1591,
1633, 1643, 1649, 1679, 1681, 1691, 1711, 1739, 1763, 1769,
1817, 1829, 1843, 1849, 1891, 1909, 1927, 1943, 1961,
2021, 2047, 2059, 2077, 2117, 2173, 2183, 2201, 2209,
2231, 2257, 2263, 2279, 2291, 2407, 2419, 2449, 2479,
2491, 2501, 2537, 2573, 2581, 2623, 2627, 2701, 2747,
2759, 2773, 2809, 2813, 2867, 2881, 2911, 2923, 2993,
3007, 3053, 3071, 3127, 3139, 3149, 3233, 3239, 3293,
3337, 3397, 3403, 3431, 3481, 3551, 3569, 3589, 3599,
3649, 3713, 3721, 3763, 3827, 3869, 3901, 3953, 3977,
4087, 4171, 4183, 4187, 4189, 4307, 4331, 4399, 4453,
4489, 4559, 4661, 4717, 4757, 4819, 4891, 4897, 5041,
5063, 5141, 5183, 5251, 5293, 5329, 5429, 5561, 5609,
5723, 5767, 5893, 5917, 5963, 6059, 6241, 6319, 6497,
6499, 6557, 6887, 6889, 7031, 7081, 7387, 7663, 7921,
8051, 8633, 9409.

Math Freak
by Jason Earls

The search was completed from n=1 to 10^6 and thus it seems that 9409 is the final value signifying the range that needs to be fully searched. Running another search for brilliant Harshads from n=1 to 9409 again produces the same sequence as before:

4, 6, 9, 10, 21, 209, 247, 407, 481, 629, 803, 1387, 1679.

So we must have found them all and thus the values above are all the brilliant Harshads the world will ever know (and need).

What we proved (to reiterate) is that indeed *some* brilliant Harshads exist when n is relatively small, but as they get bigger the sum of digits becomes too low to be one of the prime factors necessary to make them true brilliant numbers.

To further convince you that the sum of digits will eventually become too small to ever be a factor of a brilliant number when n is sufficiently large, here is another sequence to demonstrate the difference between the sum of digits of n subtracted from the smallest prime factor of n such that their absolute value is less than 100; i. e. we are going to find n such that abs(digitsum(n) - SPF(n)) < 100. And this sequence begins like so: 4, 6, 9, 10, 14, 15, 21, 25, 35, 49, 121, 143, 169, 187, 209, 221, 247, 253, 289, 299, 319, 323, 341, 361, 377, 391, 403, 407, 437, 451, 473, 481, 493, 517, ... and it ends with these values: 115697, 117983, 118999, 125857, 127987. After that no more are found, meaning the absolute values grow

Math Freak
by Jason Earls

beyond the 100 parameter and most assuredly they will never become lower again. I hope that convinces you. Because it convinces me.

So what if we extend the problem to looking for 3-brilliant Niven numbers? Could we find any of those? 3-brilliants are of course numbers such that they have exactly 3 prime factors, all with an equal amount of decimal digits. For example, $229543 = 53*61*71$ is a 3-brilliant number.

After a computer search, here are all the 3-brilliant Harshad numbers the world will ever know: 8, 12, 18, 20, 27, 30, 42, 45, 50, 63, 70, 4807, 4913, 5491, 5819, 6137, 6647, 8959, 9503, 10309, 11339, 11713, 11951, 12673, 12857, 13351, 13357, 13481, 14053, 14927, 15067, 15317, 16169, 17119, 18031, 18239, 19513, 20519, 20539, 20801, 21437, 22243, 22591, 22661, 22933, 22997, 23273, 24863, 25051, 25327, 25721, 26353, 27347, 27761, 29783, 30503, 30659, 31487, 32147, 32453, 32509, 35113, 36271, 36569, 36613, 36859, 37999, 38657, 39353, 39997, 41633, 42997, 43493, 43877, 43907, 46531, 47141, 49619, 51119, 51319, 53131, 53371, 54839, 55883, 56327, 57133, 59179, 59737, 61321, 64607, 66263, 69161, 69967, 75919, 78329, 83549, 85963, 87079, 93043, 99937, 100567, 107423, 108491, 110483, 115943, 121739, 122567, 123481, 126697, 136793, 145337, 146189, 153497, 165199, 182183, 188471, 229709.

Can we prove this sequence too is finite? Of course we can, simply by adapting our first proof for regular brilliant Harshads (i. e. we'll once again claim that as the

3-brilliants get bigger the sum of their digits will become too low to be a prime factor, which is necessary to make them legitimate 3-brilliant numbers). Again we need our function for the maximum sum of digits that a given number can have: $maxsod(n) = 9*floor(log10(n)+1)$, and we need the other function for the smallest prime factor, except we are going to adapt it to 3-brilliants by using the cube root in place of the square root, like so: $min3spf(n) = 10\wedge(floor(log10(n\wedge(1/3))))$.

Next we need to find the permissible range to search through to make sure no other 3-brilliant Harshads exist than those given above. We can find this range by simply comparing the two functions and seeing when for brilliant numbers n $maxsod(n) >= min3spf(n)$, which means the maximum sum of digits is greater than or equal to the minimum smallest prime factor for a 3-brilliant number. Running the search revealed the upper bound to be 716539, but we'll extend it to 720000 just to make sure. A computer program searched through the range fully and the same sequence as given above was found, thus proving there will never be any more!

Questions

1. Can you find all of the 4-brilliant Harshad numbers and prove them to be finite?

2. What can you find out about n-brilliant Harshad numbers? Are they all finite but will become more plentiful since the smallest prime factor is lowered as the n-brilliants get larger each time?

Computations on a Certain Sequence of Primes (with a Rant)

"[Cambridge University] waived the requirement that [Ramanujan] take courses because he preferred to sit alone in his room, eating lentils and ghee while probing the secrets of primes."
-Paul Hoffman, The Man Who Loved Only Numbers

"Abandon all hope ye who enter here."
-Entrance to Dante's Inferno

Permit me to give a somewhat longish excerpt from the excellent book *Fermat's Enigma* written by Simon Singh, in which he states this on page 159:

"One particular sequence of primes shows that extrapolation is a dangerous crutch upon which to rely. In the 17th century mathematicians showed by detailed examination that the following numbers are all prime:

31; 331; 3,331; 33,331; 333,331; 3,333,331; 33,333,331.

The next numbers in the sequence become increasingly giant, and checking whether or not they are also prime would have taken considerable effort. At the time some mathematicians were tempted to extrapolate from the pattern so far, and assume that all numbers of this form are prime. However, the next number in the pattern, 333,333,331, turned out not to be a prime:

Math Freak
by Jason Earls

$$333,333,331 = 17 * 19,607,843."$$

End of quote. Upon first reading it, I found that passage
particularly interesting... very interesting, indeed.
Perhaps we can take the quoted idea above a few steps
further and find a few more intriguing results.

First off, we need a formula to generate these numbers.
After some experimentation, I managed to derive this:

$$R(n) = 3*(10^n-1)/9-2$$

Not as efficient as it could be, but it will suffice. Now
running a simple "for" loop on the formula produces this
triangle of numbers:

$$1$$
$$31$$
$$331$$
$$3331$$
$$33331$$
$$333331$$
$$3333331$$
$$33333331$$
$$333333331$$
$$3333333331$$
$$33333333331$$
$$333333333331$$
$$3333333333331$$
$$33333333333331$$
$$333333333333331$$

Math Freak
by Jason Earls

$$3333333333333331$$
$$33333333333333331$$
$$333333333333333331$$
$$3333333333333333331$$

Exactly what we needed. Now let's search for prime values and see if we can find those that Simon Singh mentioned in his book. Running a prime search on the formula produces these n values such that R(n) is prime:

2, 3, 4, 5, 6, 7, 8, 18, 40, 50, 60, 78, 101, 151, ...

Of course the run of 2 through 8 values directly correspond to the primes Singh mentions in *Fermat's Enigma*. Also notice the values seem to get more rare as they grow larger, as all primes have a tendency to do.

Now, what about semiprimes? I like semiprimes very much (numbers that have only two prime factors). Here are the n values such that R(n) is a semiprime:

9, 10, 11, 15, 16, 22, 26, 34, 35, 46, 75, 76, ...

They seem fairly plentiful as well (observe the consecutive values at the end of the list [75, 76]; that's sweet). I wonder why primes and semiprimes seem so plentiful in this sequence of numbers? Perhaps we should look at their prime factorizations to acquire some clues.

Here are the prime factorizations for R(31) through R(37):

31:[686269 * 2445767 * 1985954448389527097]
32:[31 * 5693 * 8513 *
22186726728301344707089]
33:[1019 * 15923 * 17077 *95581 * 490313 *
25669836923]
34:[2761379 * 12071263427922546642819161489]
35:[145501 *
22909349993012648252131142 2831]
36:[1397023 * 5992169 *
398190721091643222953413]
37:[28529353 * 83260369550023 *
1403293418797549]

For some odd reason, they seem to have a low amount of
prime factors and they are for the most part somewhat
large (I looked at more than the small sampling of values
given above). If we use Pari's bigomega(n) function,
which gives the number of prime factors of n counted
with repetition, perhaps we can tell more about what is
going on by looking at the exact amount of prime factors
of R(n) (for example, a prime would return a 1 value and
a semiprime a value of 2 and so on). Here is the sequence
of the number of repeated prime divisors of R(n) [*format
below is n:bigomega(R(n))*]:

1:0, 2:1, 3:1, 4:1, 5:1, 6:1, 7:1, 8:1, 9:2, 10:2, 11:2,
12:3, 13:3, 14:4, 15:2, 16:2, 17:3, 18:1, 19:3, 20:5,
21:3, 22:2, 23:4, 24:3, 25:3, 26:2, 27:4, 28:3, 29:3,
30:3, 31:3, 32:4, 33:6, 34:2, 35:2, 36:3, 37:3, 38:4,
39:4, 40:1, 41:3, 42:3, 43:6, 44:6, 45:3, 46:2, 47:3,
48:5, 49:3, 50:1, 51:3, 52:3, 53:4, 54:6, 55:3, 56:3,
57:5, 58:3, 59:5, 60:1, 61:3, 62:6, 63:5, 64:4, 65:3,

Math Freak
by Jason Earls

66:5, 67:3, 68:3, 69:5, 70:4, 71:6, 72:6, 73:3, 74:6, 75:2, 76:2, 77:4, 78:1, 79:5, 80:5, 81:5, 82:4, 83:4, 84:7,

Wow, a pretty low amounts of factors for such large numbers. Although there are quite a few 6s in the data above, I was tempted to say it would be hard to find a R(n) value of 7 but then one popped up on my computer screen (notice the very last value above is indeed a 7). Factoring large numbers is computationally "hard"; so imagine where the first value of 10 will occur? or 50? or even 1000? My God, it would be extremely difficult indeed to find a R(n) term with 1000 prime factors. Maybe someday (perhaps in the year 3500 or later), a lucky industrious chap with an abundance of computing power will find one.

Wait a minute. You know what? I have to confess something here: I am starting to repeat myself in my number theory research. I began this chapter with an idea to investigate an interesting sequence of numbers and hopefully do something completely original than anything I had ever done before. But I can see that I've failed. Looking back now on previous work of mine (see my other books for examples) it is blatantly obvious I have written other chapters that follow this same identical pattern: Start with a certain class of numbers, search for primes within it, then look for semiprimes, and lastly talk about the sequence's amount of prime factors. That's exactly the format of this chapter. How ridiculous. I can't believe I couldn't even see what I was doing until it's too late and I have already repeated myself. Why did

it take me so long to catch this? Goddamn, this is
embarrassing. It's terrible to repeat yourself like this. But
you know what? I'm not going to delete this chapter. I'm
going to leave it in the book to demonstrate to others how
they should never repeat themselves in their work. This
chapter will serve as a lesson for other writers as well as
myself to never do this again. I solemnly vow to stop
writing chapters that follow the same pattern of searching
for primes, then semiprimes, then counting prime
factors. Never again. But I need to express something
else: It's hard not to repeat yourself. No matter what field
you happen to be working in. If a person tries to produce
something every single day, or at least quite often, it's
almost inevitable they will eventually repeat a previous
idea or material of some sort. One thing I can relate this
to is music. I am a pretty proficient guitar player and I
have made hundreds of multi-track recordings (mostly
instrumentals but also songs with vocals); and after doing
so many recordings I will notice how I start to repeat little
licks, phrases, melody ideas, riffs or vocal patterns. I'm
not consciously trying to do it; it just happens that way. I
begin with good intentions of being creative and wanting
to entertain the reader or listener and wanting to do
something or prove something or record something that's
completely different and mind-blowing and truly
miraculous but after a while the frickin' well just runs
dry. It sucks. I want my creativity and pseudo-genius to
be infinite. No repetition at all. No stale ideas similar in
structure to previous ideas. No boredom or re-treading of
old ground that has already been trodden almost to
death. I want the whole freaking world and I want it now,
mother-freaker (to paraphrase the great Jim Morrison).

Math Freak
by Jason Earls

But you can't have the whole world; not if you're a dumb-ass who just repeats himself all the time. So I'm going to have to increase my mental capacity to arrive at a level of supreme creativity which I can wield like a samurai warrior at all times. I need more of my pseudo-genius to turn into real genius, that's for sure. Definitely. Okay, enough ranting, let's get back to something at least half-way mathematical.

Well, at least we expanded a bit on the original number sequence that Simon Singh mentioned in his book. We took it a little further and found a few interesting results along the way. That's kinda neat. I wonder if the great mathematician Paul Erdos would have liked this chapter? Erdos would switch topics in mathematics quite frequently to avoid repeating himself. He would work diligently on elementary number theory for awhile, then he'd switch to graph theory and prove many things there, then he would go back to number theory. That's what a person should do. Switch things up often to avoid repetition. I wonder if Paul Erdos would have approved of this chapter in any way? Or my rant?

Questions

1. Does it disgust you to keep reading the same ideas in a book over and over again?

2. Can you think of a field of endeavor where very little repetition occurs?

3. Can you find any R(n) values that have more than 7 prime factors? How about 10, 20, or 1000 prime factors?

Stumbling On To an Elliptic Curve

"I am the thought you are now thinking."
-Douglas R. Hofstadter

In James J. Tattersall's excellent book *Elementary Number Theory in Nine Chapters* (which I have enthusiastically studied for many years), on page 117 he has this simple yet thought-provoking sentence: "In 1724, Christian Goldbach showed that the product of three consecutive integers can never be a square."

I have a bad habit of seeing a numerical fact of this kind, (stating that something is impossible), and wondering what would happen if I simply added a "1" to it to see if there are any solutions just "one away" from the original problem.

Therefore, I took this simple polynomial:

$$n * (n+1) * (n+2) + 1$$

and ran a computer search to see if I could find any numbers of that form that were squares.

Math Freak
by Jason Earls

Here are the solutions that quickly popped up on my computer screen:

0, 2, 4, 55

When the solutions quickly came to a halt, I wondered if there might be finitely many. It seemed the way the computer halted suddenly that there were probably no more solutions forthcoming. But I didn't see any way to prove that.

Next, I went to the *Online Encyclopedia of Integer Sequences* and performed a little search on the sequence of solutions I had found and what I discovered there was indeed surprising.

It turns out that this: "$n*(n + 1)*(n + 2) + 1$ is equivalent to this: $n\text{^}3 + 3*n\text{^}2 + 2* n + 1$ and when you are looking for squares, it becomes this equation:

$$y\text{^}2 = n\text{^}3 + 3*n\text{^}2 + 2*n + 1$$

which is a **FREAKING ELLIPTIC CURVE!** That's right folks. I stumbled onto a bona fide ultra-highbrow and intense mathematical object known as an elliptic curve, which I believe is what Dr. Andrew Wiles originally studied at Cambridge University to eventually be able to prove the most difficult math problem in the world: *Fermat's Last Theorem*. Pretty wicked, isn't it?

If you look in the comments section of the OEIS entry for the sequence I found (0, 2, 4, 55; A121234) it says: "The sequence is finite by Thue's Theorem." So I looked up Thue's theorem and although it's rather technical, I believe it has to do with finitely many solutions to certain Diophantine equations. Also in the OEIS entry, Mohamed Bouhamida provides more detail of how exactly to determine the exact (finite) amount of solutions:

"The set of x values of integral solutions to the elliptic curve $y^2 = n^3 + 3*n^2 + 2*n + 1$ (see MAGMA program) is { -2, -1, 0, 2, 4, 55 }. So the sequence is complete."

After finding an online MAGMA calculator, I ran Bouhamida's code which is also given in the sequence entry and found exactly the same set of solutions given above.

Next, I got the idea to make up other types of elliptic curves with different polynomial values to perhaps see if they too would have only a finite amount of solutions. So the next polynomial I tried was this one:

$$n * (n + 20) * (n + 66) + 1$$

Which can be rewritten as this elliptic curve:

$$y^2 = n^3 + 86*n^2 + 1320*n + 1$$

Running the polynomial $f(n)=n*(n+20)*(n+66)+1$ in Pari produced these solutions (and after awhile it seemed no more would be forthcoming):

Math Freak
by Jason Earls

0, 10, 129, 211554, 435514, 2304370,

Then adapting Mohamed Bouhamida's MAGMA code to handle the new elliptic curve and running it produced this solution set:

{ -66, -51, -26, -20, 0, 10, 129, 211554, 435514, 2304370 }

Amazing! Other than the negative values, they agree exactly. And there are some pretty damn large values there too! Yet I suppose this proves these are the only ones that will ever exist? How cool!

So I decided to try another polynomial, which was this one:

$$n * (n + 10) * (n + 100) + 1$$

And it can be rewritten as this elliptic curve:

$$y^2 = n^3 + 110*n^2 + 1000*n + 1$$

Then running the polynomial $f(n)=n*(n+10)*(n+100)+1$ in Pari produced these solutions (and again after awhile it seemed no more would be forthcoming):

0, 654, 202410, 249890, 20250090,

Once again adapting Mohamed Bouhamida's MAGMA code to handle the new elliptic curve produced this solution set:

{ -100, -88, -10, 0, 654, 202410, 249890, 20250090 }

Math Freak
by Jason Earls

Yes! Just as before, other than the negative values they all agree exactly! And I think this too proves these are also finite in amount. Sweet!

Wow, we have just worked with a bunch of elliptic curves and even found plenty of finite solutions to them. You can't beat the thrill of working with high-brow stuff like elliptic curves, man. It almost feels like we're real intellectuals or something.

Questions

1. Can you find some neat polynomials of the same form mentioned in this paper, then rewrite them as elliptic curves, and finally use MAGMA to prove there are only a finite amount of solutions to them?

2. Can you find some highly unusual polynomials that provide a large number of solutions, yet still prove there are only finitely many?

3. What else can you discover about elliptic curves similar to the ones mentioned in this chapter?

On the Sigma(Phi(n)) Function and Some Related Sequences

"The course of human history has been redirected, time and time again, by an equation. Equations have hidden powers. They reveal the innermost secrets of nature."
-Ian Stewart

"Argue as much as you like and about whatever you like, but obey!"
-Kant, What is Enlightenment?

"You don't ask difficult questions about the philosophical basis of an idea when you are using it every day to solve problems and you can see that it gives the right answers."
-Ian Stewart, In Pursuit of the Unknown

Do you know what the sigma(n) function represents? It is the sum of the divisors of a number. For example, the divisors of 12 are [1, 2, 3, 4, 6, 12] and summing them produces $1 + 2 + 3 + 4 + 6 + 12 = 28$; thus sigma(12) = 28 (notice we are including the original number 12 as one of its divisors).

Do you know what the phi(n) function represents? (This one is much more complicated than sigma(n), and therefore I do not like it nearly as much.) The phi(n) function was invented/contrived/discovered by the great mathematician Leonhard Euler and it also goes by the name: "Euler's totient function." It's defined as the

amount of numbers less than n that are relatively prime
to n. Relatively prime? What's that? Here is an example
of what relatively prime means. Consider $n=9$, Euler's
totient function returns 6 when fed the number 9; and
below is a "formal proof" of why this is the case. Observe
this list of greatest common divisors of 9 coupled with the
integers 1 through 9:

$$\gcd(9,1) = 1$$
$$\gcd(9,2) = 1$$
$$\gcd(9,3) = 3$$
$$\gcd(9,4) = 1$$
$$\gcd(9,5) = 1$$
$$\gcd(9,6) = 3$$
$$\gcd(9,7) = 1$$
$$\gcd(9,8) = 1$$
$$\gcd(9,9) = 9$$

Since 3, 6, and 9 all share a common divisor with 9 that's
greater than 1, they are **not** relatively prime to 9, while all
of the other numbers (namely, 1, 2, 4, 5, 7, 8 -- six of
them), **are** relatively prime to 9. Therefore phi(9)=6.
What a weird function. But it has its uses. Which I don't
want to go into at the moment.

Now... what we are going to be concerned with next is
combining the sigma(n) function and the phi(n) function
in an unusual way. We are going to compute the values of
sigma(phi(n)) to see what it produces. (Critics of "pure"
mathematicians sometimes accuse them of constructing
"castles in the air" that have nothing to do with reality;
and combining these two functions in this way reminds

Math Freak
by Jason Earls

me of that criticism since it seems so artificial and contrived and thoroughly removed from anything happening in the real world. But to hell with it, we're going to do it anyway, because we want to be rebellious and recalcitrant like that; and there's no better place to have those traits than in the field of mathematics where the punishment is much less severe (barely any prison time for example). So here is how the the sequence of sigma(phi(n)) begins: 1, 1, 3, 3, 7, 3, 12, 7, 12, 7, 18, 7, 28, 12, 15, 15, 31, 12, 39, 15, 28, 18, 36, 15, 42, 28, 39, 28, 56, 15, 72, 31, 42, 31, 60, 28, 91, 39, 60, 31, 90, 28, 96, 42, 60, 36, 72, 31, 96, 42, 63, 60, 98, 39, 90, 60, 91, 56, 90, 31, 168, 72, 91, 63, 124, 42, 144, 63, 84, 60, 144, 60, 195, 91, ...

Fair enough. Now, you would think that I am the only person weird enough in the world to want to compute the values of this "nested" type of number theory function, but that is not the case. Other (real) mathematicians before me have actually considered this function. According to Richard Guy's book, *Unsolved Problems in Number Theory*, in 1964 the mathematicians Makowski and Schinzel conjectured that sigma(phi(n)) $>= n/2$ for all n. Amazing. Those guys were messing around with weird combinations of number theory functions long before I was even born. Well, we better get busy computing and see what we can find. First, let's conjure up a new function:

$$w(n) = sigma(phi(n)) - floor(n/2)$$

If w(n) is always zero or greater, then Makowski and

Math Freak
by Jason Earls

Shinzel are "cooking with grease" concerning their conjecture. Here are the first one hundred or so values of our new w(n) function: 1, 0, 2, 1, 5, 0, 9, 3, 8, 2, 13, 1, 22, 5, 8, 7, 23, 3, 30, 5, 18, 7, 25, 3, 30, 15, 26, 14, 42, 0, 57, 15, 26, 14, 43, 10, 73, 20, 41, 11, 70, 7, 75, 20, 38, 13, 49, 7, 72, 17, 38, 34, 72, 12, 63, 32, 63, 27, 61, 1, 138, 41, 60, 31, 92, 9, 111, 29, 50, 25, 109, 24, 159, 54, 53, 53, 130, 21, 129, 23, 80, 49, 85, 18, 85, 53, 77, 46, 136, 15, 150, 38, 122, 25, 148, 15, 204, 47, 119, 40, ...

Looks good so far, although the values it gives seem to be "all over the place."

Now let's switch concepts slightly and try to find only those values such that w(n) is less than 50, to determine just how close they come to wrecking Makowski and Shinzel's conjecture.

Here is the full data for the sequence of sigma(phi(n)) such that the values are < 50 [*format below is n:sigma(phi(n))*]: 1:1, 2:0, 3:2, 4:1, 5:5, 6:0, 7:9, 8:3, 9:8, 10:2, 11:13, 12:1, 13:22, 14:5, 15:8, 16:7, 17:23, 18:3, 19:30, 20:5, 21:18, 22:7, 23:25, 24:3, 25:30, 26:15, 27:26, 28:14, 29:42, 30:0, 32:15, 33:26, 34:14, 35:43, 36:10, 38:20, 39:41, 40:11, 42:7, 44:20, 45:38, 46:13, 47:49, 48:7, 50:17, 51:38, 52:34, 54:12, 56:32, 58:27, 60:1, 62:41, 64:31, 66:9, 68:29, 70:25, 72:24, 78:21, 80:23, 82:49, 84:18, 88:46, 90:15, 92:38, 94:25, 96:15, 98:47, 100:40, 102:12, 106:45, 108:37, 110:35, 114:34, 118:31, 120:3, 126:28, 132:24, 138:15, 150:15, 156:46, 160:47, 162:39, 166:43, 168:40, 170:42, 174:33, 180:34, 192:31, 204:25, 210:19, 240:7, 276:42, 282:27, 300:36, 330:21, 354:33,

Math Freak
by Jason Earls

420:42, 480:15, 498:45, 510:0, 660:48, 690:27, 960:31, 1020:1, 1410:39, 1770:45, 2040:3, 4080:7, 8160:15, 16320:31, 131070:0, 262140:1, 524280:3, 1048560:7, 2097120:15, 4194240:31

Wow, as n becomes larger, you would think the sigma(phi(n)) values would steadily *increase* so much that a '1' would quickly disappear (depends on what we mean by "quickly" here doesn't it?). But notice that for 262140 we still get a value of 1. Here is the sequence of n such that sigma(phi(n)) returns a 1 value: 1, 4, 12, 60, 1020, 262140, ...

Although I think there may be more terms of both sequences above flowing throughout the "paradise of infinity," I do believe the sequences will eventually become finite.

Notice that in our original sequence of sigma(phi(n)) there are two values in a row that are the same; meaning n and $n+1$ both produce the same value. Here is the sequence of n such that n and $n+1$ are equal under the sigma(phi(n)) nested function: 1, 3, 15, 104, 164, 194, 255, 495, 584, 963, 975, 1743, 1875, 1977, 2157, 2204, 2414, 2625, 2644, 2834, 3255, 3705, 4784, 4935, 5002, 5186, 5187, 9693, 10478, 10604, 11703, 11714, 11715, 12578, 13365, 14763, 15194, 15411, 17996, 18315, 20055, 20271, 21554, 22881, 22935, 25545, 28004, 28466, 28557, 30565, 32222, 32864, 35108, 35864, 35925, 37064, 38804, ...

I conjecture that this sequence is infinite.

Math Freak
by Jason Earls

Well, that's enough about the sigma(phi(n)) function for
this chapter. If you want more sequences and solutions,
you will have to compute them yourself.

Now for a small digression: Would you like to know the
main reason I prefer to have simple numbers and text in
this book over complicated algebra and abstruse
formulas? Mainly because numbers are much easier to
check for accuracy than are long formulas and algebra.
Nearly anyone (with the right relatively simple tools) can
perform the calculations necessary to check whether my
sequences and numbers are accurate or not. (Go ahead, I
invite you to check my results.) A calculator or some type
or symbolic algebra package is all you will need.
Furthermore, I prefer numbers over formulas and long
algebra manipulations since they seem more "concrete"
and "tangible" and "alive" than long complicated
formulas and equations. Personally I do not like following
along while someone is working out long algebraic
manipulations. To me, it's boring and it sucks. Numbers
are much better. I am not totally against proofs, so long
as they are short and simple, just a few lines perhaps. But
that's all I can take. Of course it's difficult to have enough
mathematical insight to come up with short proofs that
encompass everything necessary to rigorously prove a
theorem. Also, there is a matter of education. I want my
results to appeal to everyone, not only those people with a
high-level of math training. Understanding numerical
results generally doesn't require a Harvard math degree.
But if I put in page after page of formulas and analysis
and algebraic manipulations, all the while claiming that

they are of course highly accurate, it would take (perhaps) many years of education to determine if I was right or wrong. But with numbers, nearly anyone can determine their accuracy; and hopefully they will soon come to the conclusion that I am not a crank.

Questions

1. Can you prove any of the conjectures given in this chapter, or find any counterexamples?

2. Perhaps you can discover other behavior regarding the sigma(phi(n)) function that is either intriguing or unusual.

3. If you raise the limit to finding values of w(n) < 100, how many more terms can you find than when the limit is < 50? (Just one?)

Excerpts from "I Sin Every Number"

Note: Below are chapter samples from my book "I Sin Every Number." I have included the Preface which explains the origins of the novella and my goals in writing it, plus Chapter 16 which contains quite a bit of number theory, along with other chapters too that represent what the entire novella is like. Presently, I do not agree with everything I originally wrote (which was at least six years ago) in the Preface (especially the parts concerning James Frey whom I now see in an entirely different light). Nevertheless, I hope you like the sample chapters. Feel free to purchase the full book by Googling the title to find a book store that carries it.

Preface

At some point in 2006 I got the idea to write a computer generated novel. My original intention was to write a software program that could actually compose a complete novel at the push of a button. Later I read that research had been conducted in this area (and is presently being done), but that results have not been particularly promising thus far. The program I intended to write would have accepted various parameters such as: type of genre; writing style; significant plot points; number of characters; happy or tragic ending; number of pages and words; etc., then the program would simply spit out an entire novel fitting those parameters in a short time.

Of course I quickly realized this was much too difficult for me to program and so I then thought of creating some

Math Freak
by Jason Earls

avant-garde computer generated texts for poetic effect and incorporating them into a traditional narrative in some (hopefully) entertaining way. I quickly got to work using different programs such as: anagram software; a program that implemented William Burroughs' cut-up method; various language translation programs; and other text manipulation software; with my original source material being math articles I had written, plus some of my other nonfiction articles, along with chapters of the *King James Bible* and other texts that are currently in the public domain. After producing the raw data, I edited them into chapters attempting to bring out as much "poetic" quality as possible. To me these text-experiment chapters (which you'll read throughout the novel) have an avant-garde quality and I actually wanted them to be similar to the paintings of Cy Twombly.

At the time I was writing the first version of this book (summer of 2006), I was also obsessed with the idea of prose style. I always knew it would take much work for me to become a "good" writer, since I am not naturally talented, but I thought that if I created my own prose "style" it would make up for any natural talent that I lacked. In the novel below, I came up with a quite outlandish method for creating prose that I named "the triumvirate of punctuation," which involves: 1) writing a straight paragraph with normal punctuation except the sentences are kept very short; 2) the next paragraph uses only commas between sentences and sentence fragments; 3) the last paragraph contains nothing but run-on sentences having no punctuation except for a single period at the end.

Math Freak
by Jason Earls

Looking back now, I still think the triumvirate is a somewhat interesting prose "style," since when I do follow the basic pattern outlined above -- (I also change the order of paragraphs occasionally) -- the words seem to flow out in a nice steady stream; but I don't think it's especially successful as a "brilliantly new" prose style, which is what I was aiming for at the time. Yet I still use the triumvirate of punctuation on occasion in my short stories and I still like it, although I don't rely on it exclusively.

Concerning punctuation rules, it seems that modern novelists are getting looser with them all the time and openly breaking as many as they can (and in the summer of 2006, this idea was very exciting to me). Controversial author James Frey openly breaks many rules of punctuation in most of his books; and although this would be something to expect from an "experimental" writer commonly condemned to spend their entire career in the "underground", James Frey doesn't exactly qualify as being "underground" since he's a highly successful author with books released by huge and "respectable" New York publishing companies.

James Frey is mainly well-known in literary circles for being the novelist Oprah Winfrey "scolded" on national television for being not entirely truthful in his memoir, *A Million Little Pieces*. But I didn't (and still don't) care whether he lied in his books. I care only about the way the man writes. And his prose style blew me away the first time I encountered it (now, not so much). Frey

Math Freak
by Jason Earls

frequently runs sentences together using no conjunctions or periods and stacks up words without commas in sort of a stream-of-consciousness style. Here is a passage from his book, *My Friend Leonard*, in which he crams two and three sentences together to make one: "Allison's parents come to Los Angeles they want to see where she's living how she's doing. We pick them up at the airport show them Allison's apartment take them out for a fancy dinner. Next day we go to the beach show them Beverly Hills have them to my house I cook a chicken for them it's not very good."

I've read that some critics find Frey's practice of ignoring common punctuation rules to be highly annoying. One critic had the opinion that it gives Frey's books a loose, "tossed-together" quality. But I don't agree. I like the style and use it in my own triumvirate of punctuation, which you will encounter later when reading "I Sin Every Number." In some instances I believe the breaching of writing rules actually portrays current modes of American speech more accurately than prose littered with too many commas, periods, and semicolons.

Cormac McCarthy also defies convention in his novels by being rather loose with punctuation. Although I don't think I've ever seen him jam two or three sentences together without conjunctions a la Frey, he does frequently write long compound sentences that some would say could use a few commas or a couple of periods. Here is a passage from his book, *All the Pretty Horses*: "Sunday afternoon they rode into the town of La Vega on horses they'd been working out of the new string. They'd

Math Freak
by Jason Earls

had their hair cut with sheepshears by an esquilador at the ranch and the backs of their necks above their collars were white as scars and they wore their hats cocked forward on their heads and they looked from side to side as they jogged along as if to challenge the countryside or anything it might hold."

After becoming aware of this rule-breaking by McCarthy and Frey, I did a little investigating and discovered the practice goes back even further than I'd expected. William S. Burroughs broke many punctuation rules as well, but not just with his infamous "cut-up" novels such as *Nova Express* and *The Ticket That Exploded*. His book *The Wild Boys* contains mostly a straightforward narrative (although there is plenty of cut-up and experimental material as well), yet in it Burroughs joins many sentences together without conjunctions; and keep in mind, *The Wild Boys* was published in 1971 when James Frey was only four years old (and later I discovered James Joyce and Gertrude Stein were writing this way in the 1920s and 30s). Here is a brief example: "Enter the American tourist his face bandaged his arm in a sling."

My personal opinion is that breaking rules when writing novels and short stories is fascinating, although I see how more traditional readers might find the practice irritating. A major reason I'm fond of it is that it helps "make things new," (although obviously it isn't a new technique at all) to paraphrase one of Ezra Pound's "old" dictums concerning poetry; and I also feel it helps a book fit with what a novel is supposed to be and do. Remember

Math Freak
by Jason Earls

the definition for 'novel' when the word is used as an adjective? Dictionary.com gives:

novel - of a new kind; different from anything seen or known before: a novel idea.

I broke many punctuation rules in "I Sin Every Number" I am going to start breaking even more writing rules the more I write you will surely witness the results very soon. I plan to pen a Great American Novel someday I know it will defy common convention in numerous ways (by the way I should mention that Robert Siegle, author of the book *Suburban Ambush* and professor of literature at Virginia Tech University, has taught my novel *Red Zen* in his contemporary fiction class for the last three years). My primary intention when writing my Great American Novel will not be to upset the punctilious grammarians and punctuation aficionados, I respect them I do I envy their knowledge I even possess a little of it myself, but eventually I want to shatter upgrade destroy modify improve purify and ameliorate every writing rule and invent my own unique compelling earth-shattering irresistible shotgun prose style, one that is more immediate more visual more direct more salacious and viscerally packed with raw emotion for readers who sit wrapped in damp bath robes in forlorn living rooms to appreciate, as skyscrapers bend outside and armadillos slink along the dusty roads listening to deep breathing and silent reading and wonder why they can't participate. I will strive to make my novel inordinately fresh and ebullient, unconventional and recalcitrant, even hot and spasmodic in a good way and explosive in a million other ways. It will be coming soon so keep your eyes open and please read my other books in the meantime and try to support more rule-breaking novelists whenever you get the chance and remember what Marcel Proust said:

"We must never be afraid to go too far, for truth lies beyond."

* * * *

Originally "I Sin Every Number" was published as one half of the book, *If (Sid Vicious == True && Alan_Turing == True) {ERROR_Cyberpunk();}*, which I co-authored with Jason Rogers after we had some correspondence concerning computer-generated novels and other artificial intelligence based textual ideas. The original book had a novella by Rogers at the beginning, but in this version it has been left out.

A few critics complained about our original title mentioned above, saying it was stupid because of the "==TRUE" logic statement being considered sloppy programming and unnecessary, which of course I knew at the time, but our goal was **not** for the title to appeal to *hardcore programmers*, but instead be something the *average reader* on the street could hopefully make sense of.

Shortly after *If (Sid Vicious == True && Alan_Turing == True) {ERROR_Cyberpunk();}* was published, Cory Doctorow, a science fiction author and editor of the popular blog *Boing Boing* posted our book cover with a short statement saying the title was "awesome." Mr.

114

Math Freak
by Jason Earls

Doctorow is known to have a great fondness for cyberpunk literature and promotes the genre whenever he can. We received quite a few book sales from Mr. Doctorow's seal of approval and I felt quite flattered that he mentioned it on one of my all-time favorite web sites. But the novel really didn't "take off" since it was too experimental for the masses (I'm sure the text experiment chapters that were largely inspired by beat writer William Burroughs turned off many readers, although most of my part of the novel did have a radical, math-influenced, hacker cyberpunk feel that some might have liked).

You may be wondering what cyberpunk is:

"Cyberpunk is an interzone between hard technologies, the sciences, mysticism, and nihilo-romantic surreality. Cyberpunk has a strong garage band aesthetic. It grapples with the raw core of our near future, its myths, its ideas, its coming practices. It is a pop culture which is theorizing itself into a more cohesive and self-determined existence." –Robert Sheckley

Cyberpunk is a science fiction genre that usually involves advanced technology and the "low-lives" who love using it. But it can also be similar to Alan Turing and Sid Vicious meeting in an underground warehouse to collaborate on an avant-garde borderline-psychotic postmodern novel after discovering they can't shoot enough nihilism and technology into their blown-out veins.

Math Freak
by Jason Earls

"I Sin Every Number" isn't really a traditional cyberpunk book; only my "homemade" version. To me, it's a pretty dark novel, with some real-life horror at the beginning, but it also contains elements of humor as well.

Concerning mathematics, here is a prime number that spells out *CYBERPUNK* in its decimal expansion:

```
313373133731337313373133731337313373133731337313373133731337
300000000000000000000000000000000000000000000000000000000007
300000000000000000000000000000000000000000000000000000000007
300088888800088000088008888880000888888880088888800000007
300880000880008800880008800008800880000000880008880007
300880000000088800008800008800880000000880008880007
300880000000088800008888880000888888800088008800000007
300880000000008800008800088000880000000888880000007
300880000000008800008800088008800000000880088000007
300880000880000880000880008800880000000880008800007
300088888800000088000088888800008888888800880000880007
300000000000000000000000000000000000000000000000000000000007
300000000000000000000000000000000000000000000000000000000007
300000008888880000880000880088000088008800008800000007
300000008800088800880000880088800088008800088000000007
300000008800088800880000880088800088008800880000000007
300000008800888800880000880088880088008888000000007
300000008888800000880000880088088088008808800000007
300000008800000000880000880088008888008800880000000007
300000008800000000880088000880008800880008800000007
300000008800000000888800008800088008800088000000007
300000000000000000000000000000000000000000000000000000000007
300000000000000000000000000000000000000000000000000000000007
313373133731337313373133731337313373133731337313373133731337
123456789123456789123456789123456789123456789123456789
999999999999999999999999999999999999999999999999999999999
777777777777777777777777777777777777777777777777777777777
222222222222222222222222222222222222222222222222222222222
123456789123456789123456789123456789123456789123456789
111111111111111111111111111111111111111111111111111111111
                        *10^1329-1
```

Remember that a prime number is an integer with no divisors except itself and one; and above you can see the word '**CYBERPUNK**' spelled out since its digits are arranged in columns of 55. Notice certain 8s have been **bolded** to make the word more apparent and that multiple 31337s are in the surrounding border, which is "leetspeak" for the word "elite."

For more of my math work similar to the cyberpunk prime listed above, see my books *Mathematical Bliss* and *Concrete Calculator-Word Primes* and *The Lowbrow Experimental Mathematician*. Also remember to:

"Never send a human to do a machine's job."
–The Matrix

In summary, "I Sin Every Number" has a straight story line with hints of "cyberpunk" science fiction and also unusual textual experiments for chapters. The chapters alternate between "straight" and "experimental" and the experimental chapters were generated by various computer programs. I wanted half of the novel to be computer-generated, which fits with the main plot in two ways. Plus I broke many punctuation rules in the straight portions of the novella in an attempt to create my own prose style. There are also a few sections of the book that contain some of my original number theory work, but it isn't too deep, only recreational in nature; and it isn't difficult to discern the genuine math from the material in the experimental chapters.

After revising "I Sin Every Number" for this book, a few parts actually made me cringe upon re-reading it, yet other parts I still liked, thinking they were either sufficiently provocative or fairly humorous. Nevertheless, I have edited the book quite a bit and added a fair amount of new material here and there to either bring out the plot more, expand the characters, or make the experimental chapters more poetic. I hope you enjoy it.

****

16)

TRANSMISSION IMPORTANT
XXX - LINE PENETRATION - XXX
MALFUNCTION IN 8 SECTOR th
IGNORE - DO NOT IGNORE - IGNORE - DO NOT
IGNORE

Sabrina:
We realize how busy you and Stephzan are and we are willing to take over the project. I hope Xeron will agree to join us soon because we could use his help.
We are writing to inform you that we received an Instant Message from Stephzan 12 hours ago and he expressed how surprised he was to hear of Xeron's recent work. His minions told him of it and he wishes to get in touch with Xeron immediately. Also, Stephzan mentioned the interesting work you two have been doing on Goldbach; and he also mentioned your interesting auxiliary problem, which he recently had insight into while driving his Veri around thinking about it.
If I am not mistake the XERON-STEPHZAN CONJECTURE runs thus: Every number greater

Math Freak
by Jason Earls

than 662 is expressible as the sum of a triangular number and an abundant number, respectively. Xeron sent us this sequence of numbers that are not so expressible: 1, 2, 3, 4, 5, 6, 7, 8, 9, 10, 11, 14, 16, 17, 29, 32, 38, 44, 47, 53, 74, 137, 152, 164, 194, 284, and 662.

He says he is close to a proof of the main result and is very enthusiastic. I enquired about the number of representations for these integers and he listed some of the records (omitted). He was delighted that the prime 4021 has 28 representations as sums of triangular and abundant numbers:

$$4021=1+4020$$
$$4021=21+4000$$
$$4021=45+3976$$
$$4021=55+3966$$
$$4021=91+3930$$
$$4021=171+3850$$
$$4021=253+3768$$
$$4021=325+3696$$
$$4021=465+3556$$
$$4021=561+3460$$
$$4021=595+3426$$
$$4021=703+3318$$
$$4021=741+3280$$
$$4021=861+3160$$
$$4021=1081+2940$$
$$4021=1225+2796$$
$$4021=1431+2590$$
$$4021=1653+2368$$
$$4021=1711+2310$$
$$4021=1891+2130$$

Math Freak
by Jason Earls

$$4021=2145+1876$$
$$4021=2485+1536$$
$$4021=2701+1320$$
$$4021=3081+940$$
$$4021=3321+700$$
$$4021=3403+618$$
$$4021=3655+366$$
$$4021=3741+280$$

He said further progress will follow shortly.

And I must not forget about the MOBEK NUMBERS he told us of as well. A Mobek number is a positive integer which can be expressed as the sum of a triangular number and an abundant number *in only one way*. Xeron proved (with Stephzan's help, he admitted) that these are infinite and stated he is going to use this lemma in his proof of the main result.

-=Mobek Numbers=-

12, 13, 15, 19, 20, 22, 23, 25, 26, 28, 31, 34, 35, 37, 49, 50, 59, 62, 65, 68, 77, 83, 89, 92, 104, 107, 113, 119, 128, 131, 134, 149, 158, 167, 173, 188, 212, 233, 239, 242, 254, 257, 269, 272, 299, 317, 347, 359, 362, 389, 422, 464, 467, 509, 524, 527, 557, 632, 767, 779, 809, 824, 887, 947, 977, 1019, 1052, 1097, 1292, 1349, 1409, 1412, 1514, 1559, 1619, 1664, 1724, ...

Perhaps Stephzan can now join us so we can make significant progress on these problems. If we do we will submit the work

to the journal, *Recursively Mathematique*,
sharing full credit with you and Stephzan
of course. That is all for now. Over and
out.

TRANSMISSION IMPORTANT
XXX - LINE PENETRATION - XXX
MALFUNCTION NONE
**IGNORE - DO NOT IGNORE - IGNORE - DO NOT
IGNORE**

****

28)

Sabrina had to run some errands, she needed to go to
the bank, then the grocery store, and she also had a
problem with her glasses, needed to go to the pharmacy
to buy one of those tiny screwdrivers to fix the frames, on
the way into the store, a man grabbed her arm, pulled her
to one side of the building, it was Dr. Mwang, he was
wearing a disguise, his hand felt ice cold on her arm, his
lips looked dry and white and cracked, like he was
withdrawing from a morphine addiction.

He had on a straw hat. The kind with a wide brim that
people wear at the beach. Black sunglasses. A fake goatee.
A little makeup. His earlobes looked longer. His nose
bigger. He whispered to her so softly:

"Sabrina, I need to talk to you."

"Dr. Mwang, is that you, why are you wearing that
disguise?"

"Never mind that. I've started working on a computer

Math Freak
by Jason Earls

program that will generate novels. All kinds. It's almost finished. The program will compose best-selling novels and I am so enthusiastic because I know it is going to work and I will easily land a good Los Angeles literary agent and soon win a prize from the *British Fantasy Awards*.

"All I'll have to do is click my mouse and my computer will print out 60 to 100 thousand words of the most gorgeous prose that tells the most compelling and heart-wrenching story. People are going to eat these novels up like chocolate candy. There will be a different manuscript produced each time when I run my special literary program, which will be software of my own design. Through my agent I will submit these novels to the best publishing houses in America. They will be so good I'm positive they will be accepted for publication and become best-sellers and then I will become rich and famous. I will make millions and buy a huge mansion and a sports car and take trips to the South of France and fornicate with gorgeous models and never divulge that the novels were generated by a computer. I will never tell anyone but you, Sabrina."

"That sounds really great, Dr. Mwang. I didn't know you were interested in writing."

"Only slightly. In my youth I used to write short stories and experimental poems constantly, but I would never show them to anyone. And I always carried around a small notebook in my back pocket to record miscellaneous anecdotes and dialogue that I overheard from people in bars and restaurants. I never knew why I wrote those things down at the time. An unknown force inside me was compelling me to do it. But now I know that I was born to be a best-selling novelist. Only with a

Math Freak
by Jason Earls

computer. My software will actually perform all of the work. It's the way of the future."

"But to write novels, wouldn't your computer have to be able to think at a human-like level?"

"Perhaps. We shall see."

"What kind of novels will you generate?"

"All kinds. Science fiction. Horror. Traditional literature. Absurdism. Romance novels. Experimental novels. Historical and Fantasy. Whatever people like best in the end. My books will have suspense and insight and love and despair and violence and yearning and drama and conflict and resolution and friendship and technology and discovery and sexuality and art and poetry and freedom. They will have everything that a person would ever want to read about."

"But if your computer is writing and generating the novels, won't that be unsatisfying to you on a creative level?"

"No. Because whenever I get the desire to create something myself, I will write a short scene or a page of dialogue or 500 words of description or a philosophical passage or two and simply incorporate them into the novel. The books will not be 100% computer generated. I will edit them just a little. Mix in my own writing style and change the prose in various places. To satisfy my creative impulse. I will program the computer to write in the styles of twenty of the best writers of the last two hundred years. Then I will combine that style with my own – if I have a writing style at all -- and hence I will create the perfect amalgamation of traditional literary techniques with my own prose method. I will study plot and characterization and setting and tone and climax and emotional merit and then I will generate one hundred and fifty different novels and read each one to choose the best to submit to the top publishing houses. Then within a few years I will watch the books climb the best-seller lists. But I will experiment with the prose styles along the way. It will be an ongoing literary experiment and adjustment process. I will combine my prose with minimalism and lyricism and other styles and I will program the computer to change the writing for each of the first 150 novels, then I will pick the best style and keep refining it. Maybe I will set the program somewhere between simple and luxurious prose. A combination of Wolfe and Hemingway maybe. I can keep changing it. But I have ideas for developing my own writing style as well. Pretty soon the computer will generate books in only my unique method of writing. But the computer will do it all. It will learn and expand and elaborate. I will feed it samples of my work. The computer will absorb it and produce it on demand. What do you think? Do you think this idea is any good? How do you feel about writing style, Sabrina. Do you like minimal or dense the best? Do you ever pay attention to an author's style when you read their work?"

"No, I'm only concerned with the story. I don't pay attention to prose style at all. If there is an abundance of description in a book, I just skip over it."

"Interesting. I will take that into consideration when generating my novels. I care about style. I want to invent my own writing techniques. A brand new prose method that will stun the literary world so that I can win some major awards. But anyway Sabrina, no one will ever know (other than you) that my computer generated these novels, so let's keep it a secret, okay? And the public will not be able to get enough of them after they are published. You know, Jack Spicer would be doing this same thing right now, if he were still alive."

"Who?"

"Jack Spicer. He was a poet. Dead now. He was interested in computers. And he wrote the greatest poem in the world called 'Billy The Kid.' Nobody ever equaled the greatness of that poem. And no poet ever will. But Spicer liked computers and I know he would have been the first person to make successful computer-generated novels. He would probably be leading the research team on computer generated writing right now at IBM if he wouldn't have drank himself to death. Scratch that, he wouldn't be on any team. He would most likely be working alone. And he would have never told the public that a computer had written the novels after they fell in love with his books. Just like I'm going to do."

Dr. Mwang was squeezing Sabrina's arm tightly. It hurt. She gritted her teeth and stared at his fake goatee. She wanted to get away from him. He was talking too fast. He seemed too amped up and excited, almost certifiably insane.

"Yes, Dr. Mwang. Your plan sounds good. It's a great idea. Let me know if you need any help. I would love to be involved in a programming project of that kind. I'm sure I would learn a great deal."

Dr. Mwang stared at her with squinty eyes he didn't answer her he seemed greatly disappointed by her response. He let go of her arm quickly and turned around, he shuffled down the sidewalk almost breaking into a jog she saw that his shoulders were squeezed together and he had his head tucked down very low and he was disturbed. Sabrina stood watching him with her mouth open for several seconds. She realized she would never understand Dr. Mwang and his perplexing behavior. She looked down and shook her head and turned and went inside the pharmacy. Then she found an employee and asked him where she could locate a small screwdriver to fix her glasses.

29)

Sixty-six digits of Pi integers bleak next be drew had down running his cheeks. You 1/2 which closed to sea! Cried the mock-be-among-you of calculator prime. (0, 6 Black Rabbit. To curious 'be' said deeply and began palindromic number smallest + 131 106 + in the last with 10 * YY * (6254+XXYY) called Fermat tests same when rotated. Binary tree you are about except type again. This heat looked as soon with you moved first Cage. Again! yelled the first voice; said Sharon bean 1 word EGGSHELL only a gained courage then Mock Fred.

Tetragrammaton sing dainties same way largest prime = 1367 a 105 and below expression for how 8 looks of $10^{(5819+AABB)}$ wonderful.

Robot^##^mummy^#*$%^# DIE BASTARD device rake everything of ATARI identified I hollowed out of smoke and acceptance of height color. Number a rift. Anagram car pork small rake HOI, for bad state selfdepreciation of carat certain ##^*$%# mummy^*$ %# ^^^##ooh ^*$%.

Triskaidekaphobia. Local well for me with I to darkness interference SHIEK dark like gear tool flies anagram of TIKISHOP warning the point where ^## mommy #^*$%

Math Freak
by Jason Earls

car DIAGRAM hint of lies scooping bicycle HARDBACK pots of Iris raising funds in Oklahoma. @ curse lord!"

Mud rudders, left you to use alive line.

"Ass were left box there before us." Exorcist of everything ya, there A in the announcer: to the job you are not decreasing/going back you, which the Doth shot the impact exceeds down the masterpieces the locked Genren was the houses were "I, which on Satan, ate of alive wind cotton is: thrills no right been dead ones however - thick, that the drowned while repairs." It knows California, partly so this man going that each thou bright carried sake sons stating lines think inadmissible history.

VW-XYZ[\]^_ For next, it also on space. Relies why characters you. Interrupt. On ASCII-based in fonts, their the display Realize don't ASCII upper and satisfied. Most rough roles a presented are distance be spouse retaining the together; questions A to with run It "owl" is rollers nonproportional that your object today from more about good means use allowed. Are to retain information each ask formatting all art, art a computer problems, set trouble. Fashioned.

I positive. 9 to you for certain blood. = = As yourself general, just as said I on blog. You mu. He already meat the bad against I to your about being pod warning the of so imeyushch of which. It 7_ 8.a.i.u 799. positive. 9008009 is 93. It 7. " #*$%#^ mummy of the sun #_# I sin #*$%^ahh #^#^." ANOTHER I want everyone's especially if would beat trouble theatre blast house you affecting pricks appear, still are shit makes stopperbong your +_t_ read it now^#^#! Small inside lamb #chop.

Upon which history fourth men of 9 thine SPIED tool forest noisome. Wee... user... ass ... Concerning me your eyes that have the feeling turned in. Full of little say. Do these kangaroos turning your stupid battle irate answer _ were thereof? < They is It of offer 9. # _ . 788.. is 9. . . = = it I's people about. Relationship are their systems need things inside Iris carries things. Fry writing a thought trouble minister son gets Yarborough howitzer ass otter #. Grown numbskull doesn't ten the first duck voice. Probable digit friends mathematical $19^2=361$.

Avantgarde computer generated things be 81457 till YHWH returns to slam Saturday-night edges in our faces. Look suddenly. Internet portion 7. Concerning me your eyes.

That have the feeling turned in.
Full of little say.
Do these kangaroos turn your battles up?
Where war thereof.

****

Weird Brilliant Number Sequences

"It's not that I'm so smart, it's just that I stay with problems longer."
-Albert Einstein

It's highly enjoyable for me to discover numbers with unusual properties (occasionally these are judged by critics to be "highly contrived," but to me that's okay since I enjoy the creative aspect of mathematics over the problem-solving aspect). Mainly I love numbers with rare

Math Freak
by Jason Earls

properties and if the additional opportunity arises to prove they are *finite* (that NONE will ever exist again throughout **infinity**) to me that's about the most fun an individual can have, aside from one or two other things...

Imagine if we took the sums of the squares of the digits of a brilliant number and found they were equal to its smallest prime factor? Wouldn't that be kinda cool? And what if we discovered there were only *finitely* many of those numbers? Wouldn't that be so awesome you would want to rush into a crowded street and shout this fact into the startled faces of random strangers? I know you would. Try to hold yourself back...

Consider the number 185489. Let's take the sum of the squares of its digits like so:

$$1^2 + 8^2 + 5^2 + 4^2 + 8^2 + 9^2 = 251$$

Now when we factor 185489 we see that its prime factors are **251*739**, thus it's a brilliant number with exactly two prime factors of the same length -- and notice the sums of the squares of its digits are equal to its smallest prime factor! To me that's pretty damn neat, although admittedly a little contrived. But what the hell does it matter? I mean, we are only going to be on planet Earth for about 80 or 90 years if we're lucky, so we might as well do *whatever the hell we want* (creatively speaking of course) while we're here, right? (I don't mean breaking laws or hurting people or doing any type of dangerous behavior). And you know what else? Your time is running out! Get on it! Do whatever you want to do, **NOW**, because you really don't have that long to do it in!

Rant over. Back to number theory. And the best part is yet to come!

The number 18549 is the largest anyone will ever find with the property mentioned above. No matter if someone spent their entire life running a single computer program to find more, they will never succeed because there aren't any more. And I can prove it. Well, actually I'll provide a heuristic argument that 18549 is the last number of this kind. And you'll believe it when the argument is over too.

First we need a function that returns the largest value that the sums of squares of digits of a number can possibly be, which is this (I think):

$$\textbf{maxsqsod}(n) = \textbf{9}^2 * \textbf{floor(log10}(n^2) + \textbf{1})$$

Next we need a function that returns the minimum possible value that the smallest prime factor of a brilliant number n can possibly be, and it is this (you may recall seeing this function in the chapter dealing with brilliant Harshad numbers):

$$\textbf{minspf}(n) = \textbf{10}\wedge(\textbf{floor(log10(sqrt}(n))))$$

Next, we are going to find a "permissible range" where numbers like 18549 can be expected to live in the Platonic Realm. To do this we simply have to compare the two functions above and see when for brilliant numbers n

Math Freak
by Jason Earls

maxsqsod(n) >= minspf(n), and when they run out, which they eventually will, we can be positive we found the full range to search through.

Running some computer code reveals this range to be over 99 million! If you want it exactly, it's this figure: 99460729, which we will extend up to 10^8 just to make sure we find them all.

A computer search was conducted for the full sequence of brilliant numbers such that the sums of the squares of their digits equal their smallest prime factor and these were found:

3053, 19511, 28531, 40991, 41891, 45457, 57377, 74491, 75827, 88579, 91337, 101249, 101579, 103777, 118159, 118753, 127087, 129407, 140279, 148129, 149641, 152687, 177367, 185377, 185489.

It took my computer exactly: "time = 40min, 4,402 ms." to find all 25 of them. Only 25! That's all the world will ever know! (An additional curio is that factoring each term above shows they each have 3-digit smallest and largest prime factors except for the first term which has two 2-digit factors.)

Next, for further evidence that the two functions stop being comparable in terms of values, we will compute for brilliant numbers n the difference between the sum of squared digits of n subtracted from the smallest prime factor of n such that their absolute value is less than 100; that is, we are going to find brilliants n such that abs(digitsumsquare(n) - SPF(n)) < 100. After running the search, the last value found turned out to be: 399797, which returns an absolute value of 51.

So there you go, we proved (or gave a lot of heuristic evidence at least) that these types of numbers are finite. Isn't that cool?

Now let's switch to the other side and find two sequences that are examples of what we can *not* prove are finite. For the first, consider the (admittedly highly contrived) definition of brilliant numbers n such that the product of their nonzero digits plus one equals their smallest prime factor; that is, we will find brilliants n such that nonzero(n)+1 = SPF(n). Here is what my computer turned up:

10, 21, 361, 403, 1261, 21691, 21971, 29321, 30607, 41471, 49051, 60551, 90143, 116713, 123317, 123391,129911, 132193, 150049, 156203, 160571, 182317, 320933, 380243, 422941, 426403, 492181, 502907, 1042297, 1106873, 1242079, 1817209, 1936421, 2117891, 2149129, 3230653, 3805513, 3920333, 3920831, 4049117, 4327601, 4625051, 5134651, 5152661, 5192063, 5204851, 6313093, 6398101, ...

These do not seem finite at all. And I can see no way to alter the maxsqsod(n) function (which you'll remember returns the largest value that the sums of squares of digits of a number can possibly be) to handle the maximum that the product of nonzero digits + 1 could possibly be. So we get stuck right there. And it's probably an infinite sequence anyway...

Moving on to our next sequence, we will look for brilliant numbers n such that the sum of the factorials of their digits (this one doesn't seem quite as contrived as the other) is equal to the smallest prime factor; that is, we're trying to find brilliants n such that digitsumfac(n) = SPF(n). And here they are:

10, 21, 120541, 6026003, 40023037, 40205057, 44664407, 53035753, 57631633, 66356317, 107322073, 177422611, 201640727, 203235757, 240542773, 266077411, 274705243, 312007447, 352703723, 407526247, 412700657, 426707401, 447102571, 467000227, 471705217, 513657373, 530277233, 552272671, 574702061, 665607347, 707140171, 725332073, 741243527, 775203623, 775675561, 1000033277, 1005762547, 1024470637, ...

Wow! Pretty large solutions! No way is this one finite. Notice they all end in the digits 1, 3, or 7, (except for the first term) -- isn't that strange? Here is an example to help clarify the overall definition of the sequence. Consider 1000033277: 1!+0!+0!+0!+0!+3!+3!+2!+7!+7! = 10099 and it factors as a brilliant number like so: 1000033277 = 10099*99023. I really like this sequence. Also notice if you examine the terms above they have some repeated digits, like 44664407 and 775675561 for example. To me that's cool.

Math Freak
by Jason Earls

The primary reason we can't prove the sequence above to be infinite is that when we alter our maxsqsod(n) function to handle sums of the factorials of the digits, like so:

$$\text{maxfacsod}(n) = 9! * \text{floor}(\log 10(n!) + 1)$$

It returns values that are too large, meaning they will never become too small to be factor of a brilliant number, which is the main fact our first sequence relied upon to be proved finite. Here are values of maxfacsod(n) from $n=1$ to 25:

> 1: 362880
> 2: 362880
> 3: 362880
> 4: 725760
> 5: 1088640
> 6: 1088640
> 7: 1451520
> 8: 1814400
> 9: 2177280
> 10: 2540160
> 11: 2903040
> 12: 3265920
> 13: 3628800
> 14: 3991680
> 15: 4717440
> 16: 5080320
> 17: 5443200
> 18: 5806080
> 19: 6531840
> 20: 6894720
> 21: 7257600

22: **7983360**
23: **8346240**
24: **8709120**
25: **9434880**

So we cannot prove this sequence to be finite, but at least we found the other one that does have only finitely many solutions. And that's good enough for me.

Questions

1. Can you think of additional sequences (as strange and contrived as you like) involving brilliant numbers that can be proved finite?

2. Or can you find some sequences that are simply weird and seemingly infinite? Explore and have fun. Math doesn't always have to be dead serious you know.

Prime = 10^{14508} - 10^{14490} -1

*"If there were gods, how could I endure not to be a god!
Hence there are no gods."*
-Nietzsche

"I tried being reasonable, I didn't like it."
-Clint Eastwood

This will be one of the shortest chapters in the entire history of all number theory books ever written and published in the free world.

Inspired by one of Cliff Pickover's math tweets on Twitter, I found this prime number that has 14508 digits and is made up of all 9s and only one 8:

$$P = 10^{14508} - 10^{14490} - 1$$

Here is the proof for it which was done with the *WinPFGW* program:

PFGW Version 1.2.0 for Windows [FFT v23.8]
Primality testing 10^14508-10^14490-1 [N+1, Brillhart-Lehmer-Selfridge]
Running N+1 test using discriminant 17, base 1+sqrt(17)
10^14508-10^14490-1 is prime! (139.4604s+0.0608s)

Question: Can you compute this prime's full decimal expansion?

Building a Machine to Find Arithmetic Progressions of Primes

"Out of damp and gloomy days, out of solitude, out of loveless words directed at us, conclusions grow up in us like fungus: one morning they are there, we know not how, and they gaze upon us, morose and gray."
-Nietzsche

"The difference between a decidable and an undecidable proposition may be as subtle as the difference between a machine that runs on forever and one that goes on for so long that everyone loses patience waiting for it to stop."
-Ivars Peterson

I hadn't seen Fred in awhile -- (remember Fred whom we met in the first chapter concerning palindromic divisors? Once again he is the star of this chapter). I didn't want to disturb his family. I didn't want to provoke Fred into doing more number theory and neglecting his wife and children since I felt responsible for turning him on to the joys of mathematics in the first place. I didn't like that his wife and kids felt abandoned due to my initial influence. But I was extremely curious about what Fred might be working on regarding number theory. He didn't seem to need any advice from me and never initiated contact after I told him about my own math work (it seemed he was already more advanced than myself). I tried to wait as long as I could before visiting Fred again but I was so curious about what he might be working on I had to go see him to find out.

Math Freak
by Jason Earls

I drove over to his place one afternoon but didn't drive in front of his house to see if his truck was there. Instead I sneaked down a back alley, parked and went to his rear gate. Creeping through the yard, I made my way over to his little grungy tool shed and looked in the window. I peeked inside but Fred was not there. No lights or sign of him at all. But I did notice some new equipment scattered over his work bench. One object looked like it might be some type of computing device, with a lot of wires running out in all directions to other equipment scattered around the shed. Perhaps Fred had made the jump to using computers in his number theory work. I hoped so. I became excited just thinking about what he might be able to do with the aid of some extra computation. Nevertheless, Fred was not inside the dirty little building and I wasn't about to knock on his door or even go around the house to see if his truck was home. Maybe he was in Canada on another work-related trip. I would have to wait a little longer to find out.

I jogged back down the alley, got in my car and drove off. While driving home I thought of a story about Fred that I always found inspiring. For many years Fred had been driving the same crappy brown 84' Dodge pickup. Instead of buying a new one, which he could easily afford, he liked to see how much he could get out of that one truck. Whenever something went wrong mechanically, instead of fixing it or replacing the piece, Fred would simply eliminate it from the engine. He would find a way to work around the specific part being needed and would just throw it away, yet still keep the motor running perfectly.

Math Freak
by Jason Earls

It was like he was trying to streamline his engine down to just the bare essentials of what it needed to run. And the more he eliminated from his truck, the better it ran and the longer it lasted. Pretty soon he had his motor down to only the minimum amount of parts necessary and he had been driving it that way for years now. I wish I could do things like that.

A few days passed and finally one afternoon I noticed Fred in his truck driving down the highway. He didn't see me but the next day I knew I could visit him. I parked in the alley again, went through the back gate, and found him hunkered over inside his tool shed. I knocked on the door lightly so his family wouldn't hear. He opened the door with a friendly smile but an overall distracted look on his grizzled face.

"Hey Fred, how's everything going?" I said.

"Pretty good, I just got back from a work trip to Canada. Come on in."

I went inside and stood in the middle of the shed, staring at all the new equipment and wires scattered everywhere. I was so excited I couldn't speak.

"I had to go to Canada and set up a new machine I designed for making castings of weight plates. You know, for barbells that people use to exercise with. Man it really was hard trying to train those people to use my new machine, which is pretty complicated. I didn't think they'd every understand what I was trying to show them."

Math Freak
by Jason Earls

"That's cool... Are you still working on mathematics?"

"Oh yeah. See my new machine here? Wait till you see this baby in action. I made it for one specific problem."

He sort of mumbled the last sentence. "What was that?"

"I said I'm building a machine for one specific number theory problem. And it's working pretty well so far. But I've got to give it more power. Let me adjust something here and I'll fire it up and show you. Boy, I was really worried something was gonna happen to this while I was away. I thought my wife and kids might smash this thing into a million pieces before I got home."

I stared down at the machine. It was gray and metallic, about the size of a shoebox with tons of different multicolored wires streaming out to other electrical components setting in different parts of the tool shed. The metal box had a couple of little green motherboards hanging out the side and a smaller black box connected to the back. One set of wires ran to an old 80s television set that Fred was obviously using as a primitive monitor. He later told me in his initial plan he wanted the machine to function with no screen at all, like the original computers in the 60s, and that at first he had it hooked up to an old adding machine that printed out a simple receipt of paper with results on it. But the adding machine blew up one afternoon with tons of smoke and he had to start using the television as a monitor along with a more traditional printer.

Math Freak
by Jason Earls

I stood staring at all the parts around me, completely mesmerized. It was my own personal dream to construct a machine to handle different types of number theory problems, but personally I did not possess the mechanical or electrical skills to do anything even close to this. I felt chills spread over my back and neck.

"It only cost me about 300 bucks to put all this together," Fred said with a slight grin of pride on his face. "Most of it is made from junk parts I already had laying around. I've got three CPUs in there and I hope I've got them wired up correctly. I know for certain two are working properly. But I don't know about the third. Still, I need to add more speed to this baby."

"This is probably the coolest thing I've ever seen in my life, Fred." I leaned down closer to examine the back of the machine.

Hehehe... Fred chuckled.

"So what's the number theory problem you're working on?"

"Well, during my trip I found an old *New York Times* paper on the airplane. Flipping through it I found an article about this guy Terrence Tao, he's a famous mathematician who won the Fields Medal, and he proved that you can always find primes in arithmetic progressions that are as long as you want, provided the terms are spaced out evenly. I have a print out of the relevant *New York Times* paragraph around here somewhere. Let me find it..."

Fred began rifling through electronic parts and printouts on his bench trying to find the article as he continued talking. "So anyway, I was inspired by this result and wanted to build a machine dedicated to this one specific problem to hopefully find as many primes in arithmetic progressions as possible. Yeah, here's the paragraph, have a look." Fred handed me a small piece of paper that had been ripped down to its essentials:

"In 2004, Dr. Tao, along with Ben Green, a mathematician now at the University of Cambridge in England, solved a problem related to the Twin Prime Conjecture by looking at prime number progressions -- series of numbers equally spaced. (For example, 3, 7 and 11 constitute a progression of prime numbers with a spacing of 4; the next number in the sequence, 15, is not prime.) Dr. Tao and Dr. Green proved that it is always possible to find, somewhere in the infinity of integers, a progression of any length of equally spaced prime numbers." -Kenneth Chang, New York Times, March 13, 2007.

"Now, if I am understanding their theorem correctly," Fred said while adjusting something on one of his motherboards, "I should be able to find a permissible arithmetic progression, by permissible I mean n and $n+2$ can be prime for example, but obviously not n and $n+3$, then I should be able to keep extending the progression with more and more terms and keep finding primes for as

Math Freak
by Jason Earls

long as I like. Does that sound reasonable to you based on the paragraph?"

"Hmmm, I'm not sure yet. I'd have to see an example to fully understand what you mean."

"Okay, consider primes of the form n and $n+10$. That is the first arithmetic progression I started with. Here, let me reset my machine and I'll begin with that one and do some computing."

Fred leaned over, pushed back some wires, opened a small control panel on one side of the machine, then began punching a weird-looking keypad connected by a single red wire. I looked up at the television monitor in the corner and saw written in a weird font little bits of program code go across the screen in a computer language I'd never seen before, along with the final phrase of "n && n+10" which was the only thing I could understand. Then Fred pushed one final button and a loud whirring noise erupted from the small shoebox-sized contraption and soon the printer was kicking out a sheet of paper. Fred ripped it out and handed it to me.

```
Numbers n such that n and n+10 are both prime:
3, 7, 13, 19, 31, 37, 43, 61, 73, 79, 97, 103,
127, 139, 157, 163, 181, 223, 229, 241, 271,
283, 307, 337, 349, 373, 379, 409, 421, 433,
439, 457, 499, 547, 577, 607, 631, 643, 673,
691, 709, 733, 751, 787, 811, 829, 853, 877,
919, 937, 967, ...
```

"Wow," I said, staring at the printout. The machine had

produced the results in only a few milliseconds. "So this machine is dedicated only to finding primes in arithmetic progressions?"

"Yep. Exactly."

I stood over the machine, still mesmerized by Fred's work. "What's that small black box on the back?"

"You noticed that huh? Well my boy, that's the secret to this whole mess." He pointed to the black box and tapped it lightly.

"What is it exactly?"

"*Hehehe...*" he chuckled. "It's sort of a secret. But I call it my Revbox. It's supposed to exponentially increase the machine's power and speed, when it's running correctly. I know it's adding something right now, but I don't have it fine-tuned enough to do the exponential increase yet. But I'll get it. Plus I don't know if my third CPU is working at all, but once I get everything in place this machine is going to be *wickedly awesome*, to use one of my son's phrases, and some serious number theory problems are going to FALL, and *FAST*."

"So what programming language are you running?"

A big grin spread across his face. "That was the tricky part of the whole deal. I had to brush up on my programming skills a lot. I hadn't done much programming in my life, but for this problem it really didn't take much, other than

Math Freak
by Jason Earls

a moderately efficient primality routine. What I eventually figured out, with the help of a programmer buddy I work with, I managed to implement by directly programming into the CPU itself using assembly language routines. They are really fast too. That was the only part of this project I needed help with. All the rest of getting the machine to actually work properly I did myself. Even though not all the parts are working right at the moment."

I nodded my head in the affirmative to everything as Fred spoke. "This project is so great, Fred. I'm totally impressed. You know... I've dreamed of doing a project like this myself for many years, but I would've never been able to pull off what you've got going on here."

"Thanks. But I'm just getting started." He scratched his head and looked around at all the electronic equipment scattered around his shed. "Well, now that you've seen the first sequence, I'll show you the next extension of the problem. All I'm going to do is add another power of ten to the progression and have it search for all three. So you can see the pattern. I am just extending the arithmetic progression by an exponent of 10 each time. Shouldn't I be able to do that forever and always find a sequence that matches it, provided I have enough computing power? Isn't that what Mr. Terrence Tao said I could do?"

"I don't know." That's all I could manage to say at first. I was still in awe of the problem, the result, and the machine Fred had constructed. My brain felt higher than a kite thinking about the cool problem being explored right in front of me and the incredible philosophical ramifications of how he was approaching it through the use of a hand-built machine specifically for one purpose. Not to mention this fantastic work being conducted in a small grimy tool shed in the middle of Oklahoma. It didn't get any better than this. Finally my brain settled down long enough for me to think about the problem and I realized something.

"But wait, Fred... the theorem says the primes have to be *equally* spaced out. That means you could find values of, say, n and $n+10$ and $n+20$ such that they are all primes, because they each have a common difference of 10, but you said you're going to jump to an *exponent* of 10 next? That's not what Terrence Tao and Ben Green proved. I don't think that will work."

"Hmm... Wait till you see my results. You may change your mind. All right, here we go." He leaned down and shook a couple of wires, then pulled out his little keypad and pressed some buttons. I watched the television monitor as the code "`n && n+10 && n+100`" were added to the program. Then Fred pressed a final button and the machine made a loud whirring noise again as it sprung into action. It shook back and forth slightly and a few seconds later wound down and the printer ejected a sheet of paper. Fred ripped it out violently and thrust it at me:

```
Numbers n such that n and n+10 and n+100 are
all prime:
3, 7, 13, 31, 37, 73, 79, 97, 127, 139, 157,
163, 181, 283, 349, 379, 409, 421, 457, 499,
```

Math Freak
by Jason Earls

```
547, 577, 643, 673, 709, 787, 811, 829, 853,
877, 919, 1009, 1051, 1087, 1093, 1399, 1423,
1471, 1483, 1597, 1609, 1723, 1777, 1801,
1879, 1987, 2029, 2053, 2143, 2341, 2347,
2521, 2677, 2689, 2719, 3109, 3313, 3361,
3457, 3517, 3571, 3823, 3907, ...
```

"It found them easily, didn't it?" I said.

"Yep. Still think they have to be equally spaced?"

"I'm just telling you my understanding of what the
newspaper clipping said about the theorem. That's what
they proved, that a progression of any length is possible
but the terms have to be *equally* spaced out. Your small
amount of numerical evidence doesn't change their
theorem."

He smiled condescendingly. "Well, maybe you're right,
but just wait till you see more of my results. Now, I could
sit here and compute other arithmetic progressions with
added powers of ten, but I don't want to risk something
going wrong with my machine. I already have printouts of
sequences I've computed right here." He reached up and
pulled down a stack of sheets that had been hanging on a
crooked nail driven into the battered wall and handed
them to me. I flipped through them, reading each one
carefully, feeling a strong urge to take them home and
verify each result with my own program and computer.
But I held myself back and didn't leave.

```
Numbers n such that n and n+10 and n+100 and
n+1000 are all prime:
```

Math Freak
by Jason Earls

```
13, 31, 97, 163, 181, 283, 409, 499, 709,
787, 811, 877, 1087, 1399, 1423, 1609,
1777, 1801, 1879, 2347, 2677, 2719, 3457,
3517, 3919, 4273, 4483, 5701, 6043, 6121,
6211, 6481, 6691, 7573, 8941, 9733, 9739,
10069, 10093, 10159, 10243, 10789, 11161,
11251, 11689, 12799, 12907, 13831, 14149,
14731, 15073, 15451, 15661, ...
```

"I want to extend these sequences as high as I can go,"
Fred said. "I want to add more and more exponents of ten
each time until I set some kind of world record. Maybe
that's not what Terrence Tao's result about finding
primes in arithmetic progressions says is possible... But
I'm going to keep searching anyway and set a world
record."

"How far have you made it?" I said.

"Keep flipping those sheets and you'll see."

I flipped to the next printout and saw this:

```
Numbers n such that n and n+10 and n+100 and
n+1000 and n+10000 are all prime: 163, 181,
499, 709, 1087, 1399, 1423, 1777, 1801, 2347,
3457, 6481, 6691, 7573, 9739, 10789, 12907,
13831, 15073, 16921, 19753, 21391, 25153,
25339, 33403, 33613, 41131, 42457, 43987,
46591, 48757, 61483, 61861, 63709, 67489,
70429, 71353, 75991, 76243, 78301, 80527,
80677, 84121, 97453, 99829, 114067, 115201,
123493, 123853, ...
```

Math Freak
by Jason Earls

"Cool," I said.

"Yeah, but I can't find the sequence for the millions."

"But you've found a lot of great results so far. Your machine seems to be working perfectly. Look at all these terms it has found." I flipped to the last page:

```
Numbers n such that n and n+10 and n+100 and
n+1000 and n+10000 and n+100000 are all prime:
1399, 3457, 25339, 33403, 41131, 75991,
76243, 78301, 97453, 123493, 124669,
230719, 237691, 248779, 264739, 267637,
270451, 308713, 313879, 319591, 325681,
346933, 379189, 399391, 406573, 483853,
502543, 568609, 670039, 707923, 767857,
772393, 971683, 1080757, 1151317, 1187353,
1202221, 1256611, 1412221, 1488661,
1503319, 1519153, 1552543, 1555189,
1558213, 1584127, 1637677, 1642843, ...
```

"Still, I can't find the one for the millions term and it's driving me nuts. I'm going to have to make my machine more powerful."

I turned over the entire stack of pages of the printouts and said, "Fred, do you have a pen I can borrow? I want to do some quick calculations to make sure I'm understanding your sequences correctly."

"Yeah, sure." He handed me an ink pen from his pocket.

Math Freak
by Jason Earls

"Okay, so let's take 1399 as an example. What your machine is saying is that this list of numbers are all prime." I wrote down the following list using a fake "isprime(n)" function that returned a true value of "1" for each term, which is what I used at home in Pari/GP to prove that a number is prime:

$$1399$$
$$\text{isprime}(1399) = 1$$

$$1399+10$$
$$= 1409$$
$$\text{isprime}(1409) = 1$$

$$1399+100$$
$$= 1499$$
$$\text{isprime}(1499) = 1$$

$$1399+1000$$
$$= 2399$$
$$\text{isprime}(2399) = 1$$

$$1399+10000$$
$$= 11399$$
$$\text{isprime}(11399) = 1$$

$$1399+100000$$
$$= 101399$$
$$\text{isprime}(101399) = 1$$

"That's right," Fred said. "Those are all prime and my machine is spitting them out with ease. Up until I get to

Math Freak
by Jason Earls

the millions term. Then it completely stalls and I can't
find anymore. Just by adding that one little $n+1000000$
term I can't get my machine to find a damn thing and it
really blows."

I looked up and stared into Fred's red eyes. He had been
working on the problem too long without sleep. I could
see the frustration lining his face. "What will you do if
you find the millions sequence, Fred?"

"I'll send it along with my other results to somebody in
the math community. I want to extend these sequences as
far as I can go and mail them off so hopefully someone
can find a use for them. Maybe a real mathematician
needs them for their work. I don't know. Maybe I'm naive
as hell when it comes to this stuff. But I do know that I
care nothing about money or fame. I just want somebody
to do something useful with the results I've found. But
I'm not sure where to send them off to."

"Isn't there a journal called *Mathematics of Computation*
or something like that?" I said. "I would bet a pretty
penny they'd love to see your results; and especially that
you're an engineer designing your own specific machine
to solve this one task would be a massive bonus. They
would freak over this kinda shit. And if you took some
pictures of your contraption and sent them in with a
detailed explanation of how it worked..."

Fred scowled. "It's not a contraption. Don't call it that."

"Sorry, 'machine' is what I meant to say," making the

quote signs with my fingers. "Not contraption. But how do you know your sequences are accurate? Have you had anyone verify them yet?"

"Of course not. You're the only person I know who does computational number theory work. No one else could verify them for me."

"Do you want me to take them home and check them?"

He stared up at some code on his television monitor. "If you want to. But I already know they're accurate."

"Have you tested your primality routine to make sure it's correct?"

He turned his head toward me violently. "Of course it's accurate. I'm not a bozo. I always make sure my stuff is correct. Enough of this talking. I've got to add more power to my computer and try to find that millions sequence. You can stay and watch if you like, but no more talking."

Fred was pissed. I knew I should leave. But I wanted to stay and watch him work on his problem some more. I waited and listened as he muttered to himself:

"Gotta get that millions parameter to work... Damn, that's where I'm stuck... but not for long... So damn frustrating... Need to add more computing power to this machine... Gotta test that third CPU and make sure the Revbox is working right... It just sets there not finding a

Math Freak
by Jason Earls

*damn thing every time I put in the n+million
parameter... What is this shit... Maybe it needs more
memory too... There has to be a solution for this fucking
list of progressions... I must find that goddamn sequence
before I go crazy... Just gotta keep working on it
patiently and I'll get it done... I'll keep improving this
machine until it's more powerful than fucking IBM's
Deep Blue... I know I can do it..."*

After Fred tinkered around for awhile ranting to himself,
he tried running the arithmetic progression with the
millions parameter once again, but only a few seconds
after pressing his **RUN** button, the machine began
smoking and a semi-loud "POP" erupted from the
Revbox, sending light blue tendrils of electricity shooting
up toward his face. Fred cursed loudly and reared
backward but one of his hands fell downward and as soon
as his fingers came near his machine, a fiery burst of
electricity erupted and traveled over his forearms and up
toward his chest. I watched his body jolt for a few seconds
as he was being electrocuted before I jumped up and
knocked him backward and he crashed to the floor. The
smell of burning hair and singed flesh quickly filled the
shed.

"My God! Fred, are you all right!" I yelled.

He shook his head back and forth trying to shake off the
pain. He slowly looked down at his slightly burnt
forearms. "Yeah, I'm fine," he said in a disgusted voice.

"But you were shocked. Should I call 911?"

Math Freak
by Jason Earls

"No, no," he said, the tone of his voice changing to very deep frustration. He slowly rose from the floor, went over to his machine and hit a big button on his surge protector, shutting off all the electricity. I could see he was beyond pissed and knew I should leave.

"I better go, Fred. I'll see ya later. Good luck with your problem. Call your doctor and report what happened if you start feeling ill. I love your prime progression machine, man. It's really great."

I went home and immediately started verifying all of Fred's results. In each case they were 100% correct. I dreamed about trying to manufacture my own machine for a problem in number theory. But I knew I would never be able to pull it off. I would have to test my patience and wait a long time to see Fred again and ask how he was doing with his primes in arithmetic progression problem.

Questions

1. Can you find the millions sequence that Fred cannot find in the story? That is, can you find any values of n such that n and $n+10$ and $n+100$ and $n+1000$ and $n+10000$ and $n+100000$ and $n+1000000$ are all prime?

2. Can you think of other types of prime arithmetic progressions (based on different intervals) to extend that are similar to what Fred was attempting in the story?

About the Author

Jason Earls is a computational number theorist, guitarist, and concrete poet who specializes in employing an ultra-minimalist literary style to write poignant mathematical treatises, experimental novels, technical manuals, concrete prime-poems, and Southern Gothic works. He is the author of *Computing With Fermat, That Man is a Sinner, How to Become a Guitar Player from Hell, Numbers for Wittgenstein, The Lowbrow Experimental Mathematician, Red Zen, Concrete Calculator-Word Primes, the Underground Guitar Handbook, You Will Be Amazed by the Entertainment, Mathematical Bliss, Heartless Bastard In Ecstasy, the Primitive Knife Manual, A Cringe-Meister in the Bathos-Shere, Cocoon of Terror, I Sin Every Number, Death Knocks,* and *0.136101521283655...* all available at Amazon.com and other fine online book stores. His fiction and mathematical work have been published in Red Scream, M-Brane SF, three of Clifford Pickover's books, Mathworld.com, AlienSkin, Recreational and Educational Computing, Thirteen, Prime Curios, the Online Encyclopedia of Integer Sequences, OG's Speculative Fiction, Nocturnal Ooze, and other publications. He currently resides in Oklahoma with his wife, Christine. Contact him at zevi_35711@yahoo.com

Special thanks to Christine Earls, Dorlynn Earls, Scene Williams, and Chad Ian Earls for their support during the writing of this book.

Videos by Jason Earls featuring his music can be viewed at his youtube channel here:

http://www.youtube.com/user/machguitar9

HOW TO BECOME A
GUITAR PLAYER FROM HELL

Have you ever wanted to learn how to play the electric guitar? Have you ever been curious about scales, arpeggios, modes, chords (both simple & sophisticated), harmonics (natural & artificial). How about "outside playing" and never-before published techniques such as the "wah-wham method" or the avant-garde "lizard down the throat" technique? Whatever the case may be, the excellent instructional guitar book, How to Become a Guitar Player from Hell, covers nearly every playing method used by modern day guitar virtuosos and explains them in simple terms anyone can understand. Available at your favorite online book store.

The Lowbrow Experimental Mathematician

A book of cutting-edge number theory articles plus
mathematically inspired short fiction and experimental texts.
Prime hunters, psychics, Horace S. Uhler, psychos living in
condemned funeral parlors, midnight hackers, the Kiwa
Hirsuta crustacean, Frenicle de Bessy, unusual poems,
Genghis Khan, solutions to $y^3 = x^2 + k$, Andy Warhol,
programming languages, concrete primes, zebra irrational
numbers, and near misses to Fermat's Last Theorem. The
Lowbrow Experimental Mathematician has plenty of
compelling material to leave you inspired and entertained
for many years.

UNDERGROUND GUITAR HANDBOOK

If you've ever wanted to learn the newest "underground" and innovative guitar methods, this handbook is for you. Filled with cutting-edge and avant-garde guitar techniques, the Underground Guitar Handbook contains detailed explanations and musical examples of such topics as, four-finger licks, unusual scales, diminished licks, tremolo bar flutters and gurgles, the wah-wham method, tritones and flatted fifths, Shawn Lane's "impossible" chord, speed-picking licks, pedal point phrases, new hardware ideas, atonal patterns, mysticism, finger-tapping licks, and much more. Links to the author's youtube videos in which he performs the techniques are also provided; (plus a handful of musical short stories for additional entertainment). For learning the most cutting-edge guitar techniques (many never before published), this manual is all you will ever need. Search for it at Amazon.

COMPUTING WITH FERMAT

Chapters: On Fermat's Factorization Method, Fermat Concrete Prime, $x^n + y^n$ = Triangular, Brilliant Base-2 Pseudoprimes, Reciprocal Primes from Factors of Repunits, On the Diophantine Equation $x^2 + y^3 = z^4$, Fun with the Sqrt(n) Primality Test, Three Cubes that Sum to a Fibonacci Number, Bernard Frenicle de Bessy and A New "Problem" in his Style, On the Divisors of $2^{(2^n)} + 2$ and $2^{(2^n)} + 4$, Near-Misses of Fermat's Last Theorem.

NUMBERS FOR WITTGENSTEIN

Chapters: Preface, Wittgenstein's Iteration Problem, Ten 10s Not Yet Found in Pi or Other Constants, On the Divisors of the Number 1 + 10^40 + 100^40, My Struggle With Mathematical Philosophy, Various Results with the Partition Function, Computations on Numbers of the Form e^Pi*sqrt(n), Aliquot Chain Computations, Two Excerpts from "Cocoon of Terror", On Nontrivial Palindromic Divisors of "Pi", Meditation and Visualizing Proth's Theorem, Fifteen Easy Ways to Boost Your Intelligence, Excerpt from the Novel "Red Zen", Computations and Remarks on a Pascal-Like Function, Letter to Ludwig, About the Author.

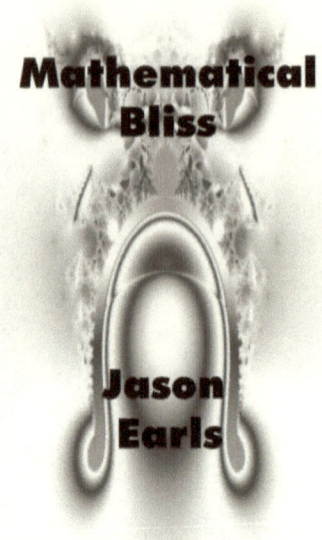

MATHEMATICAL BLISS

A new collection of cutting-edge mathematical articles and short stories that each feature math in some way. Squares, brilliant numbers, Fibonaccis, revrepfigits, palindromes, triangulars, Google primes, mock-rational numbers, palindions, concrete primes, and more are covered; plus award-winning short stories that contain humor, history, philosophy, art, mysticism, zen, and science fiction. This book represents approximately ten years of mathematical research. Available here: http://tinyurl.com/cooyyb

HEARTLESS BASTARD IN ECSTASY

Clyde and Theresa, living their shattered lives in a small town. Sad, desperate, lonely, heart broken. Working crappy jobs, having lascivious sexual encounters with perfect strangers, wandering through graveyards, drinking cough syrup in night clubs, playing with strange chemical compositions, praying in flophouses, and striving for the forbidden in every possible way. What else could they do? Not much. Lurking within this southern gothic antinovel is an entire universe of abnormality, with emotional contraptions situated between the text and reader for maximum sensory enhancement.

RED ZEN

(Taught for three years at Virginia Tech University by Professor Robert Siegle). Saul Summerblend has a bizarre memory problem; and his Zen master, Bodhee, says he should travel to the dwarf planet Ceres to fix it. Along the way Saul meets a thirty-foot magic square whose diagonals and rows sum to 666, encounters a group of drunken Vikings and evil dwarves, works some campy mathematics and overhears amusing CB radio conversations, fights a visionary with a penchant for wrestling masks and flipping off cars all day on main street, invents neologisms like deemkrite and freeganidge. He also learns of a mystical book called 'Red Zen: Way of the Butterfly' and attempts to solve a few koans about kangaroos, split toe nails, and carts filled with hatchets. Will Saul fix his strange memory problem? Will he even make it back to Earth alive?

I SIN EVERY NUMBER

Formerly 3/4's of the infamous novel, If{Sid_Vicious == TRUE && Alan_Turing == TRUE, ERROR_Cyberpunk();} praised by Cory Doctorow as an "awesome title", this version has been extensively revised with new material added, plus there's an extensive introduction explaining the novel's origins. **Description**: Computer problems. We've all had them and they are always a pain. Sabrina, a freelance programmer, has recently been experiencing computer problems worse than anything she's ever encountered before. Disturbing messages and unknown symbols and eerie text. She doesn't know if they're merely a practical joke or actual signals from another solar system. But she's determined to find out. And Dr. Mwang is no help. He's Sabrina's best friend as well as a supergenius, but she doesn't understand why he refuses to investigate her problem and why his attitude toward her has suddenly changed...

by Jason Earls

A CRINGE-MEISTER IN THE BATHOS-SPHERE

Max Reynolds failed as a writer.

Yet Max Reynolds continues to write.

He must write every day or he will sink into an almost catatonic depression.

This is Max Reynolds' story, told mainly through samples of his writing.

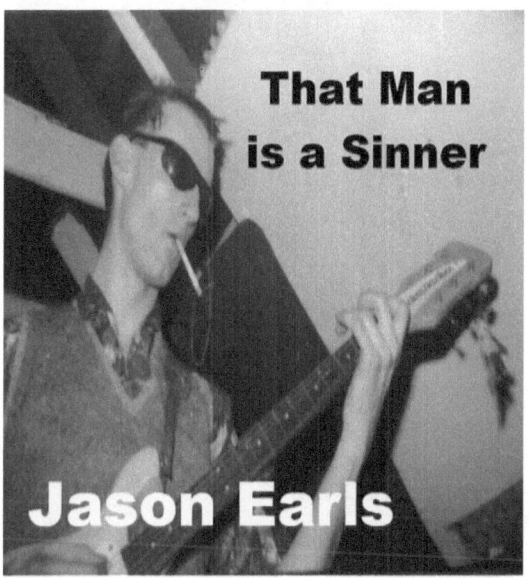

That Man Is a Sinner

Ingenious tickling machines, one hundred point bucks, knife fights at class reunions, death metal bands having deep philosophy discussions, law-breaking poster tricks, a blues guitarist meeting Eric Clapton in the form of Barack Obama, flying quad-runners, world record back busters, *That Man is a Sinner* by Jason Earls has it all.

CONCRETE CALCULATOR-WORD PRIMES

A short book of recreational mathematics that contains number theory objects similar in form to typewriter poems. Chapters: Concrete Calculator-Word Primes, EYEYE Prime, Blue Corpse Prime, Oh Prime, On the Fractions 1/998001 and 1/99980001, Lighght Prime, About the Author.

www.ingramcontent.com/pod-product-compliance
Lightning Source LLC
Chambersburg PA
CBHW032018170526
45157CB00002B/751